NOW 2 KNOW!

Algebra 2 & Trigonometry

by T. G. D'Alberto

Pithy Professor Publishing Company

Brighton, CO

Published by

Pithy Professor Publishing Company, LLC
PO Box 33824
Northglenn, CO 80233

ISBN: 978-0-9882054-6-8

Library of Congress Control Number: 2014953608

Printed in the United States of America

About the Author

Dr. Tiffanie G. D'Alberto has a Ph.D. in Electrical & Computer Engineering from Cornell University and a B.S. and M.S. in Electrical Engineering from Virginia Polytechnic Institute & State University.

She has worked for over a decade in the telecommunications and aerospace industries as a scientist, program manager, and supervisor. She has engaged in numerous opportunities for tutoring, teaching, and mentoring throughout her career and schooling.

In her spare time, Tiffanie enjoys oil painting, drawing, reading, sewing, and running. She's a huge fan of Star Trek, Renaissance Festivals, and animals.

Tiffanie lives in St. Croix with her fiancé, Colin, and their many wonderful pets.

Dedication

To my dearest Colin, who inspires me, encourages me, and supports me. I could never thank you enough.

To my high school Trigonometry teacher, Mr. Klima, who made learning math easy

Acknowledgements

I always thank my family first: My parents for the foundation, the push, and the belief in me all along; My fiancé for his inspiration, encouragement, and unending support.

A huge thanks goes to my many excellent math teachers from middle school to high school to college that not only taught the material but also taught the way of thinking necessary to excel in these subjects.

I'd also like to acknowledge Barron's Review Course Series **Let's Review: Algebra 2/Trigonometry** by Bruce C. Waldner, M. A. and **Calculus with Analytical Geometry**, Second Alternate Edition by Earl W. Swokowski for a thorough review of the material.

Finally, I'd like to thank Amazon.com for their excellent publish-on-demand service that enables books such as these, and you, the reader, for making this investment in your future.

Table of Contents

Introduction   **1**

    Welcome!   1

    Layout   2

Notation   **3**

PART 1: NUMBERS & FUNCTIONS   **5**

Chapter 1: Algebra I Review   **7**

    Types of Numbers   7

    Fractions   8

    Exponents   9

    Order of Operations & Properties   10

    Working with Binomials   11

Chapter 2: Complex Numbers   **13**

    The Imaginary Number   13

    Complex Numbers   14

    The Quadratic Equation   15

Chapter 3: Functions   **17**

    Functions   17

    Inverse Functions   19

Chapter 4: Logarithms & Exponentials — 21

General Logarithms & Exponentials — 21

Natural Logarithms & Exponentials — 22

Properties of Logarithms & Exponentials — 23

Factorials — 25

PART 2: GRAPHING — 27

Chapter 5: Graphing Overview — 29

Graphing - A Review — 29

Some Common Functions — 30

Chapter 6: Lines — 33

Distance, Midpoint, & Slope — 33

Equations for Lines — 34

Chapter 7: Circles & Conics — 39

Circles — 39

Parabolas — 40

Ellipses — 41

Hyperbolas — 43

Circle & Conic Summary — 45

Chapter 8: Systems of Equations   **47**

Systems of Independent Equations   47

Combination   48

Substitution   49

Graphing Solutions   50

Chapter 9: Congruent Transformations   **51**

Translations   51

Reflections About the Axes   54

Rotations   56

PART 3: Trigonometry   **59**

Chapter 10: Trigonometric Functions   **61**

Sine, Cosine, & Tangent   61

Cosecant, Secant, & Cotangent   65

Chapter 11: Inverse Trigonometric Functions   **67**

Arc-Functions   67

Solving Trigonometric Equations   69

Chapter 12: Trigonometric Identities   **71**

Pythagorean Identities   71

Angle Sum & Difference Identities   73

Double & Half Angle Identities   74

Product & Factoring Identities   75

Chapter 13: Laws of Sines & Cosines   77

Law of Sines   77

Law of Cosines   79

Chapter 14: Hyperbolic Trig Functions   81

Euler's Identity   81

Hyperbolic Trig Functions   82

Inverse Hyperbolic Trig Functions   84

Appendices   85

Appendix A: Course Summary   86

Appendix B: Problem Sets   88

Appendix C: Solutions to Problem Sets   104

Appendix D: Trigonometric Identities   146

Appendix E: Hyperbolic Trig Identities   147

Index   148

Introduction

Welcome!

This text is separated into three parts. Algebra 2 has two main components, analytical & graphing, which correspond to the first two parts of this text. You'll learn about imaginary numbers, logarithms, and exponentials, as well as how to sketch a hyperbola and translate and rotate objects. Part 3 covers Trigonometry: the study of relationships in triangles that is even more versatile than what you learned in Geometry.

The process of learning any math is three-fold:

1. **To excel at math is to understand math.** For example, you know how to play Go Fish. You not only know the rules, you understand the object of the game and the techniques that are required to dominate against your 4-year old opponent. It doesn't matter that this time you have a different hand or that's it's been 10 years since you last played. You *understand* the game, so you can play it well. That's how you should learn math.

2. **To understand math, you need the story** . The story is the logic flow that allows you to keep building on your *understanding*. If someone tells you a story and skips a critical part of the plot, you would and should say, "Hey, back up!"

3. **To understand math, you also need the big picture.** The big picture is the outline of the logic, or story, placed in an area small enough for you to see it in its entirety. Like a file directory on a computer, it organizes the information. Once you see the flow of the big picture, it's easier for you to put the details of the story into their proper places.

The key to learning math is not memorization, it's understanding. Be open to changing the way you think. Once you get the flow, you'll get the A's. I wish you great success!

Layout:

The layout of this text is different from most academic books:

1. **The problem sets are saved to the end of the book.** In this book, you can read from beginning to end to understand the logical progression of the course, or stop to do problem sets as you desire.

2. **Solution sets give the critical steps to get the answers, not just the answers.** Because this is not a textbook for a classroom, there is no need to keep the "secret sauce" from you. Use the problems as drills or study the solutions as further examples.

3. **Appendix A is an overall summary of the entire book.** It helps you visualize the big picture and logic flow to give you a framework into which you can organize the details.

In addition, the following visual markers will help you navigate the material...

Key terms defined for the first time are **bolded** and also found in the index.

Important equations are shown as:

> *important information*

Finally, examples are given as supplements to the text as well as for illustration:

Illustrative graphics and additional notes are shown on the side to accompany the text.

> *Example* This is an example to illustrate a point or to give further definition. Skip it if you feel very comfortable with the material presented thus far.

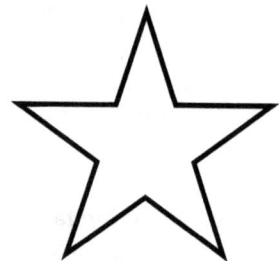

Notation

The following notation is often helpful when talking about Math. Some of this notation will be used in the text.

Closed interval – includes points a and b: [a,b]

Open interval – up to but not including points a and b: (a,b)

$\exists$	There exists
Я or R	With respect to
$\forall$	For all; For each; For every
$\ni$	Such that
$\in$	Is an element of
\	Except
iff or $\Leftrightarrow$	If and only if
b/c	Because
b/w	Between
f^n	Function
⦾	On the order of (think 3 O's)
$\rightarrow$	Implies, Is given by
$\therefore$	Therefore
α	Proportional to
$\sim, \approx, \cong$	Equivalent to; Approximately equal to
$\equiv$	Is assigned to; Is defined as; Is forced to equal to
$\neq$	Is not equal to
$\parallel$	Parallel to
$\perp$	Perpendicular to; Orthogonal to
$\cup$	Union
$\cap$	Intersection

Final Note: variables are usually u, v, w, x, y, z, and constants are usually a, b, c, and k

Part 1: Numbers & Functions

Chapter 1: Algebra I Review

Types of Numbers:

Let's start with some quick review. There are many types of numbers, some of which are outlined below:

Real Numbers: Any number between $-\infty$ and $+\infty$.
Positive Numbers: Numbers greater than 0.
Negative Numbers: Numbers less than 0.
Integers or **Whole Numbers**: Numbers with nothing after the decimal point.
Rational Numbers: Numbers that are equal to an integer divided by a non-zero integer.
Irrational Numbers: Numbers where the digits after the decimal go on forever without a pattern.
Prime Numbers: An integer greater than 1 that is NOT evenly divisible by integers other than itself and 1.
Composite Numbers: Integers greater than 1 that are not prime.

When working with negative numbers, the following table is helpful:

	$+$	$-$
$+$	$+$	$-$
$-$	$-$	$+$

To use the table for $+/-$: Find the column and row that match sign with the operator and operand to find the final operator e.g. $+(-1) \rightarrow -1$.
To use the table with $\times/\div$: Find the sign of the two operands to get the sign of the answer: e.g. $-1 \times +1 = -1$.

Example

Evaluate:

$3 - (+2)$ $3 - (-2)$ $(-3) \times (+2)$ $(-3) \times (-2)$

$3 - 2 = 1$ $3 + 2 = 5$ $-3(2) = -6$ $-3(-2) = 6$

Fractions:

Fractions are another way of showing division. There are several ways to work with fractions. The most useful way is to remember that you can always modify a fraction by multiplying top and bottom by the same number. This is true because it is equivalent to multiplying by 1:

$$\frac{2\times3}{3\times3} = \frac{2}{3} \times \frac{3}{3} = \frac{2}{3} \times 1.$$

Reduction: To **reduce** a fraction, you simplify it by dividing **numerator** (top) and **denominator** (bottom) by all common factors.

Reduce $\frac{16}{36}$: $\quad \frac{16}{36} = \frac{16\div2}{36\div2} = \frac{8}{6} = \frac{8\div2}{6\div2} = \frac{4}{3}$

Rationalization: When you have a **radical** (root) in the denominator, you can **rationalize** (get rid of it) by multiplying numerator and denominator by that radical.

Rationalize $\frac{5}{\sqrt{2}}$: $\quad \frac{5}{\sqrt{2}} = \frac{5\times\sqrt{2}}{\sqrt{2}\times\sqrt{2}} = \frac{5\sqrt{2}}{2}$

Addition & Subtraction: To add (or subtract) fractions, change top and bottom of each fraction so they both have the **least common denominator** (LCD) – the smallest number that is evenly divisible by both denominators, then add (or subtract) numerators.

Subtract $\frac{2}{3} - \frac{1}{4}$: $\quad \frac{2\times4}{3\times4} - \frac{1\times3}{4\times3} = \frac{8}{12} - \frac{3}{12} = \frac{8-3}{12} = \frac{5}{12}$

Multiplication & Division: To multiply (or divide) fractions, multiply (or divide) the numerators and multiply (or divide) the denominators.

Multiply $\frac{2}{3} \times \frac{1}{4}$: $\quad \frac{2\times1}{3\times4} = \frac{2}{12} = \frac{1}{6}$

Exponents:

An **exponent** is a way of getting a number (or **base**) to operate on itself. It is also called raising a number to a **power**, and it is denoted by a superscript. There are several types of exponents:

Positive Integer Exponents	Base should be multiplied by itself exponent times	$2^4 = 2 \times 2 \times 2 \times 2 = 16$
Negative Integer Exponents	Base should be multiplied by itself exponent times then inverted	$2^{-4} = \frac{1}{2 \times 2 \times 2 \times 2} = \frac{1}{16}$
"0" Exponent	Any base to the 0 power is 1	$2^0 = 1$
Fractional Exponents	Treat numerator as integer exponent; the denominator takes a root of the base	$4^{3/2} = \sqrt{4 \times 4 \times 4}$ $= \sqrt{64} = 8$

Addition & Subtraction: Perform each exponent operation, then add or subtract.

$$2^2 - 3^2 = (2 \times 2) - (3 \times 3) = 4 - 9 = -5$$

Multiplication & Division: If the bases are the same,
- Multiplication is achieved by adding exponents;
- Division is achieved by subtracting exponents

$$2^2 \times 2^4 = 2^{2+4} = 2^6 = 64$$

Raising to a Power: A number with an exponent can have another exponent applied to it by multiplying the two exponents:

$$(2^2)^3 = 2^{2 \times 3} = 2^6 = 64$$

Absolute Value: An absolute value applied to a number with an exponent applies to the base only:

$$|(-2)^{-4}| = |-2|^{-4} = 2^{-4} = \frac{1}{2 \times 2 \times 2 \times 2} = \frac{1}{16}$$

Order of Operations & Properties:

The **order of operations** when evaluating an expression is:
- Parentheses & Absolute Values
- Exponents
- Multiplication & Division
- Addition & Subtraction

Note: Parentheses are implied when there are multiple elements in the numerator and denominator :

$$\frac{2+4}{3+2} = (2+4) \div (3+2)$$

Example

Evaluate $4 + 3(2+4) - 5(|3-2|)$

$$4 + 3(2+4) - 5(|3-2|)$$
$$= 4 + 3(6) - 5(|-1|)$$
$$= 4 + 3(6) - 5(1)$$
$$= 4 + 18 - 5 = 17$$

The **properties of addition and multiplication** are:

Commutative	Order doesn't matter	$2 + 3 = 3 + 2$
Associative	Grouping doesn't matter	$2 + (1+3) = (2+1) + 3$ $= (2+3) + 1$
Distributive	Multiply through parentheses	$2 \times (4+3) = 2 \times 4 + 2 \times 3$
Identity	The identity turns a number into itself	$2 + 0 = 2; \ 2 \times 1 = 2$
Inverse	The inverse turns a number into its identity	$2 + (-2) = 0$ $2 \times 2^{-1} = \frac{2}{2} = 1$

The **additive identity** is 0, the **multiplicative identity** is 1.

Working with Binomials:

To multiply two binomials, use **FOIL**:

FOIL = First + Outer + Inner + Last

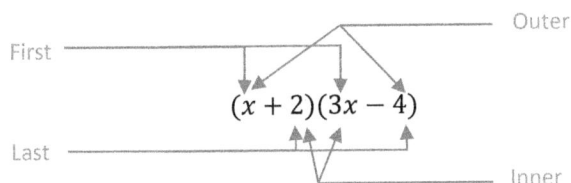

First — Outer

$$(x + 2)(3x - 4)$$

Last — Inner

The solution is: $3x^2 - 4x + 6x - 8 = 3x^2 + 2x - 8$

To go backward and factor an expression $ax^2 + bx + c$ into two binomials use the following rules:
- **a** comes from **FIRST**
- **b** comes from **OUTER + INNER**
- **c** comes from **LAST**

$ax^2 + bx + c = (\ x +\ \ )(\ x +\ \ )$	Factors for a and c ADD to get $\|b\|$
$ax^2 - bx + c = (\ x -\ \ )(\ x -\ \ )$	
$ax^2 \pm bx - c = (\ x +\ \ )(\ x -\ \ )$	Factors for a and c SUBTRACT to get $\|b\|$; the larger product gets sign in front of b.

> Factor the following: $x^2 - 5x + 6$
>
> *Example*
>
> What combination of factors of a and c ADD (there is a '+' in front of c) for $\|b\| = 5$? $(1 \times 3) + (1 \times 2)$. The answer is then $(x - 3)(x - 2)$.

Of course, you can always use the **quadratic equation** for not so easily factored expressions. If $ax^2 + bx + c = 0$, then

$$x = \frac{-b \pm \sqrt{b^2 - 4ac}}{2a}$$

Chapter 2: Complex Numbers

The Imaginary Number:

If you find yourself needing to use the quadratic equation but come up with a negative number under the square root, never fear. Someone has already imagined a way to deal with that.

The **imaginary unit**, or **imaginary number**, i, is defined as:

$$i = \sqrt{-1}$$

Now when you end up with a negative number in the square root, you can express it in terms of i:

$$\sqrt{-4} = 2\sqrt{-1} = 2i$$
$$\sqrt{-20} = 2\sqrt{(-1)(5)} = 2i\sqrt{5}$$

There are a few rules when dealing with i:

$i = \sqrt{-1}$	$\frac{1}{i} = -i$
$i^2 = -1$	$\frac{1}{i^2} = \frac{1}{-1} = -1$
$i^3 = -i$	$\frac{1}{i^3} = \frac{1}{-i} = i$
$i^4 = 1$	$\frac{1}{i^4} = \frac{1}{1} = 1$

Lets look at some examples.

Example

i^7	$\dfrac{6i}{3i^3}$	$5i(2i^2)$
$= i^4(i^3)$	$= \dfrac{6}{3i^2}$	$= 10i^3$
$= (1)(-i) = -i$	$= \dfrac{6}{3(-1)} = -2$	$= -10i$

Complex Numbers:

When real and imaginary numbers are mixed together, we have what is called a **complex number**. The general form of a complex number is $a + bi$.

$$4 + 2i = 4 + 2\sqrt{-1}; \quad a = 4, b = 2$$

There are two new operations that stem from dealing with complex numbers:

Special Set Notation:

$\mathbb{R}$ - Real numbers
$\mathbb{C}$ - Complex numbers
$\mathbb{Z}$ - Integers
$\mathbb{N}$ - Positive integers
$\mathbb{Q}$ - Rational numbers
$\emptyset$ - Empty set: { }, not {0}

To find the real part of a complex number:
$$\mathfrak{Re}\{a + bi\} = a$$

To find the imaginary part of a complex number:
$$\mathfrak{Im}\{a + bi\} = b$$

Example

Find the real and imaginary components of $5 - 3i$.

$$\mathfrak{Re}\{5 - 3i\} = 5$$
$$\mathfrak{Im}\{5 - 3i\} = -3$$

As for other operations, to add/subtract complex numbers, group like numbers together. In other words, add/subtract the real numbers and the imaginary numbers separately, much like you would do if you were working with a polynomial. Similarly, to multiply complex numbers, use the FOIL method as you would with polynomials.

Example

Evaluate $(4 - 2i) - (3 + i)$ and $(4 - 2i)(3 + i)$.

$$(4 - 2i) - (3 + i) = (4 - 3) + (-2i - i) = 1 - 3i$$

$$(4 - 2i)(3 + i) = 12 + 4i - 6i - 2i^2$$
$$= 12 - 2i + 2 = 14 - 2i$$

To divide complex numbers, you must rationalize the denominator by multiplying top and bottom by the **complex conjugate, ()***, of the denominator. The complex conjugate of $a + bi$ is simply $a - bi$ (change sign on the i coefficient). We write $(a + bi)^* = a - bi$.

Example

Evaluate $(4 - 2i)/(3 + i)$.

$$\frac{(4-2i)}{(3+i)} = \frac{4-2i}{3+i}\left(\frac{3-i}{3-i}\right) = \frac{12-4i-6i+2i^2}{9-3i+3i-i^2} = \frac{12-2-10i}{9+1+0i}$$

$$= \frac{10-10i}{10} = \frac{10}{10} - \frac{10i}{10} = 1 - i.$$

The Quadratic Equation:

The **quadratic equation** finds the factors, or **roots**, of a **quadratic polynomial** (having **order** 2, i.e. 2 is the largest exponent). The roots are where the polynomial crosses the x-axis when it is plotted.

Now that we have complex numbers, we can analyze the quadratic equation even for polynomials without real roots. Saying that a polynomial has no real roots means there are no real numbers for which it crosses the x-axis.

Example

Find the roots of $x^2 - 2x + 5$.

$$x = \frac{-b \pm \sqrt{b^2 - 4ac}}{2a} = \frac{2 \pm \sqrt{4 - 4(1)(5)}}{2(1)}$$

$$= \frac{2 \pm \sqrt{-16}}{2} = \frac{2 \pm 4i}{2}$$

$$x = (1 + 2i), (1 - 2i)$$

Example

Find the roots of $x^2 + x + 1$.

$$x = \frac{-b \pm \sqrt{b^2 - 4ac}}{2a} = \frac{-1 \pm \sqrt{1 - 4(1)(1)}}{2(1)}$$

$$= \frac{-1 \pm \sqrt{-3}}{2} = \frac{-1 \pm i\sqrt{3}}{2}$$

$$x = \left(-\frac{1}{2} + \frac{i\sqrt{3}}{2}\right), \left(-\frac{1}{2} - \frac{i\sqrt{3}}{2}\right)$$

It is also possible to work backward to recover the polynomial when you know the roots. Simply use FOIL.

Example

Find the polynomial whose roots are
$$(1 - i), (1 + i).$$

$x = (1 - i), (1 + i)$, so we have two factors:

$$(x - 1 + i) = 0; \ (x - 1 - i) = 0.$$

Multiplying them together:

$$\left(x + (-1 + i)\right)\left(x + (-1 - i)\right)$$

$$= x^2 + [(-1 - i)x + (-1 + i)x]$$
$$+ (-1 + i)(-1 - i)$$

$$= x^2 - 2x - ix + ix + (1 + i - i - i^2)$$

$$= x^2 - 2x + (1 + 1)$$

$$= x^2 - 2x + 2$$

Chapter 3: Functions

Functions:

You may have heard the word "function" and thought it was another name for an expression with variables. However, a **function**, denoted $f(x)$, has a special definition:

> $f(x)$ **is a function iff**
> **for every** x **value there exists no more than one** y **value.**

In other words, functions must pass the **vertical line test** – if you sweep a vertical line across the x-axis, the function should only intersect that line once at any given time. The notation $y = f(x)$ means that y is a function of x.

Example

These are functions.

These are NOT functions.

Example

Is this expression a function? $y = 4x + 2$

There appears to be only one y value for every x. You may recognize this as an equation for a line, which definitely has a 1:1 correspondence between x and y. We can say then $y = f(x) = 4x + 2$.

$y = 4x + 2$.

Example

Is this expression a function? $y^4 = 4x^2$

Reducing the equation, we have $y = \pm\sqrt{2x}$.

There is clearly more than one y value that would result from a given x value. For example, if $x = 2$, $y = +2$ and $y = -2$ are valid solutions. This is not a function.

$y^4 = 4x^2$.

A **composite function** is a function of a function written as:

$$(g \circ f)(x) = g(f(x))$$

Where $g(x)$ and $f(x)$ are each functions. In the above equation, you plug in the expression for $f(x)$ everywhere you see an x in $g(x)$. Note that $(f \circ g)(x)$ and $(g \circ f)(x)$ aren't necessarily the same.

Example

Find $(g \circ f)(x)$ and $(f \circ g)(x)$ if

$$g(x) = 4x^2 \text{ and } f(x) = 2x + 5.$$

For $(g \circ f)(x)$, we plug in the entire expression of $f(x)$ everywhere we see an x in $g(x)$:

$$(g \circ f)(x) = 4 \cdot (2x + 5)^2 = 16x^2 + 80x + 100$$

For $(f \circ g)(x)$, we plug in the entire expression of $g(x)$ everywhere we see an x in $f(x)$:

$$(f \circ g)(x) = 2(4x^2) + 5 = 8x^2 + 5$$

Example

Find $(g \circ f)(x)$ and $(f \circ g)(x)$ if

$$g(x) = 3x + 2 \text{ and } f(x) = x^2 + 2x + 1.$$

$$(g \circ f)(x) = 3(x^2 + 2x + 1) + 2$$

$$= 3x^2 + 6x + 5$$

$$(f \circ g)(x) = (3x + 2)^2 + 2(3x + 2) + 1$$

$$= 9x^2 + 12x + 4 + 6x + 4 + 1 = 9x^2 + 18x + 9$$

Inverse Functions:

An **inverse function**, denoted $f^{-1}(x)$, undoes what a function does. Just like subtracting a number from itself gives the additive identity, 0, and multiplying a number by its inverse gives the multiplicative identity, 1, an inverse function operating on its function gives x: $f^{-1}(f(x)) = x$.

> *Example*
>
> Is $g(x)$ the inverse of $f(x)$ and vice versa if
> $$g(x) = 4x^2 \text{ and } f(x) = \frac{\sqrt{x}}{2}.$$
>
> We evaluate $(g \circ f)(x)$ first:
> $$(g \circ f)(x) = 4\left(\frac{\sqrt{x}}{2}\right)^2 = 4\left(\frac{x}{4}\right) = x$$
>
> We can say $g(x) = f^{-1}(x)$.
>
> We evaluate $(f \circ g)(x)$ next:
> $$(f \circ g)(x) = \frac{1}{2}\sqrt{4x^2} = \pm\frac{2x}{2} = \pm x$$
>
> If we limit our answers to $+x$, then $f(x) = g^{-1}(x)$.

To find an inverse of a function, switch the x's and y's (remember $y = f(x)$), and solve for the new y.

> *Example*
>
> Find the inverse of $f(x) = 2x^3 + 4$.
>
> We rewrite $y = 2x^3 + 4$ as $x = 2y^3 + 4$,
>
> and solve for y: $y = \sqrt[3]{(x-4)/2}$.
>
> Check:
> $$f(y) = 2\left(\sqrt[3]{\frac{x-4}{2}}\right)^3 + 4$$
> $$= \frac{2(x-4)}{2} + 4 = x - 4 + 4 = x$$

19

Chapter 4: Logarithms & Exponentials

General Logarithms & Exponentials:

The **logarithm**, or **log** for short, of a number is given by:

$$\log_b y = x$$

where b is a constant called the **base** of the log. When $b = 10$, it is usually not written explicitly. In English, the equation asks to what power, x, does b have to be raised to give y.

The inverse of a log is the **exponential** given by:

$$y = b^x$$

This equation asks what y is once b is raised to the x power. One way to remember the relationship between the log and the exponential is to picture b getting on the log, sailing across the river, and hoisting x on its shoulders.

$$\log_b y = x$$

$$y \underset{b}{=\!=\!=} x$$

$$y == b^x$$

One way to remember how to perform a log function is to picture the logarithm as a log and the equal sign as a river.

Example

$$\log 100 = x$$
$$100 = 10^x$$
$$x = 2$$

$$\log_2 8 = x$$
$$8 = 2^x$$
$$x = 3$$

Because the exponential is the inverse of the log and vice versa, they are **inverse functions** of each other.

$$f^{-1}(\log_b x) = b^x; \quad f^{-1}(b^x) = \log_b x$$

If you perform an inverse function on a function, you will get the argument, x, as the answer:

$$b^{(\log_b x)} = x; \quad \log_b(b^x) = x$$

Natural Logarithms & Exponentials:

The special case of the logarithm where $b = e \approx 2.71828$ is called the **natural logarithm** and is given the symbol, **ln**.

$$\log_e x = \ln x$$

The inverse of the natural log is the **natural exponential** given by e^x or **exp(x)**:

$$f^{-1}(\ln x) = e^x = \exp(x); \quad f^{-1}(e^x) = \ln x$$

And, by definition of an inverse function we write:

$$e^{(\ln x)} = x; \quad \ln(e^x) = x$$
$$x > 0; \quad \quad for\ all\ x$$

Note that the argument of the natural log must be greater than 0. This is because $\ln(x)$ is only defined where $x > 0$.

This all may seem rather arbitrary to consider a special case where the base is equal to some weird value and call it natural. However, the natural log has a formal definition which you will learn in Calculus I.

Example

Solve for x: $\ln x^3 = \ln 64$

Taking the inverse:
$$e^{\ln x^3} = e^{\ln 64}$$

$$x^3 = 64$$

$$x = 4$$

Properties of Logarithms & Exponentials:

Logs and exponents have special properties that make them surprisingly user-friendly. For example, multiplication and division get replaced with addition and subtraction. The following table lists the laws of logs. The second column shows the special case of the natural log.

$$\log_b(p)^r = r\log_b p \qquad\qquad \ln(p)^r = r\ln p$$

$$\log_b(pq) = \log_b p + \log_b q \qquad\qquad \ln(pq) = \ln p + \ln q$$

$$\log_b\left(\frac{p}{q}\right) = \log_b p - \log_b q \qquad\qquad \ln\left(\frac{p}{q}\right) = \ln p - \ln q$$

$$p, q > 0$$

Notice that the logarithms in each relation have the same base. To change bases, the following formula is helpful:

$$\log_b x = \frac{\log_a x}{\log_a b}$$

Exponentials obey the same laws as any expression with exponents that you have worked with thus far. The following table summarizes those laws along with a second column for the natural exponential.

$$(a^p)^r = a^{pr} \qquad\qquad (e^p)^r = e^{pr}$$

$$a^p a^q = a^{p+q} \qquad\qquad e^p e^q = e^{p+q}$$

$$\frac{a^p}{a^q} = a^{p-q} \qquad\qquad \frac{e^p}{e^q} = e^{p-q}$$

$$(ab)^p = a^p b^p$$

Let's look at some examples.

Example

Solve for x: $3 \ln x = \ln 8$

$$\ln x^3 = \ln 8$$

$$e^{\ln x^3} = e^{\ln 8}$$

$$x^3 = 8 \rightarrow x = 2$$

Example

Solve for x: $8(2^{3x}) = 4^{2x}$

Try to find like bases:

$$2^3(2^{3x}) = (2^2)^{2x}$$

$$2^{3x+3} = 2^{4x}$$

$$3x + 3 = 4x \rightarrow x = 3$$

Example

Solve for x: $9^{x-5} = \frac{1}{27^x}$

Try to find like bases:

$$(3^2)^{x-5} = (3^{-3})^x$$

$$3^{2x-10} = 3^{-3x}$$

$$2x - 10 = -3x \rightarrow x = 2$$

Example

Solve for x: $3^{2x} = 5$

Assuming the table of log properties will save us, take the log (base 10) or the natural log of both sides since they can be evaluated on a calculator:

$$\ln 3^{2x} = \ln 5$$

$$2x \ln 3 = \ln 5$$

$$x = \frac{1}{2}\left(\frac{\ln 5}{\ln 3}\right) = \frac{1}{2}\left(\frac{1.61}{1.10}\right) = 0.73$$

Solve for x: $2^x = 3^{x-1}$

$$\ln 2^x = \ln 3^{x-1}$$

$$x \ln 2 = (x - 1) \ln 3$$

$$x(\ln 2 - \ln 3) = -\ln 3$$

$$x = -\frac{\ln 3}{\ln\left(\frac{2}{3}\right)} = \frac{-1.10}{-4.05} = 2.71$$

Evaluate: $\log_6 10$

You can't use a calculator to take a log of base 6, so use the conversion formula:

$$\log_6 10 = \frac{\ln 10}{\ln 6} = \frac{2.30}{1.79} = 1.29$$

Check: $6^{1.29} = 10.09$ which is correct with rounding error.

Evaluate: $\log_{1/3} 20$

You can't use a calculator to take a log of base 1/3, so use the conversion formula:

$$\log_{1/3} 20 = \frac{\ln 20}{\ln 1/3} = \frac{3.00}{-1.10} = -2.73$$

Check: $3^{-(-2.73)} = 20.07$ which is correct with rounding error.

Factorials:

The **factorial** of a positive integer is that number times one less of itself, times one less of that, times one less of that, and so on until you get to 1. The factorial is represented by an exclamation point after the number. For example:

$$5! = 5 \times 4 \times 3 \times 2 \times 1 = 120.$$

By definition, $0! = 1$.

If you happen to be without your calculator and need to estimate the natural exponential of a number, you can use the following approximation:

$$e^x = 1 + x + \frac{x^2}{2!} + \frac{x^3}{3!} + \frac{x^4}{4!} + \cdots$$

The more terms you use, the better the approximation.

Example

How many terms are needed to approximate e^2 to two decimal places?

We keep adding terms to three decimal places until there is no change in the second decimal place:

Term	Value	Sum
1st	1	1.000
2nd	$x = 2$	3.000
3rd	$\frac{x^2}{2!} = \frac{4}{2} = 2$	5.000
4th	$\frac{x^3}{3!} = \frac{8}{3\times2\times1} = \frac{8}{6} = 1.3333$	6.333
5th	$\frac{x^4}{4!} = \frac{16}{4\times3\times2\times1} = \frac{16}{24} = 0.6667$	7.000
6th	$\frac{x^5}{5!} = \frac{32}{5\times4\times3\times2\times1} = \frac{32}{120} = 0.2667$	7.267
7th	$\frac{x^6}{6!} = \frac{64}{6\times120} = \frac{64}{720} = 0.0889$	7.356
8th	$\frac{x^7}{7!} = \frac{128}{7\times720} = \frac{128}{5040} = 0.02540$	7.381
9th	$\frac{x^8}{8!} = \frac{256}{8\times5040} = \frac{256}{40320} = 0.0064$	7.387
10th	$\frac{x^9}{9!} = \frac{512}{9\times40320} = \frac{512}{362880} = 0.0014$	7.388

Ten terms were needed to confirm an answer of 7.39. By comparison with a calculator, $e^2 = 7.3891$.

Part 2: Graphing

Chapter 5: Graphing Overview

Graphing – A Review:

To **graph** a solution to an equation, we do the following:
1. Collect ordered pairs by choosing a few x values and calculating the resulting y values.
2. Draw a horizontal number line that represents the x values. This is called the **x-axis**.
3. Draw a vertical number line that represents the y values. This is called the **y-axis**. The crossing of the axes is the **origin** at $(0,0)$.
4. Plot the ordered pairs, i.e. put a dot at the **abscissa** (x value) and **ordinate** (y value) of each ordered pair.
5. Draw a smooth curve or line connecting the dots.

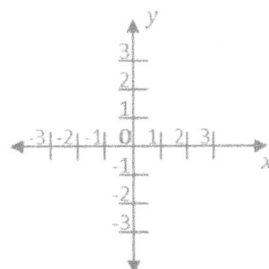

The x and y number lines for a graph. The axes are labeled and the double arrows indicate that each number line goes from $-\infty$ to ∞.

There are a few more terms associated with graphing that you should know. The **independent variable** is the one we get to pick values for at random. Usually, x is the independent variable.

The **dependent variable** takes on values dictated by the equation and the values for the independent variable. The dependent variable, y usually, is the one isolated to the left of the equation ($y = \cdots$).

The span of x values is called the **domain**, and the resulting span of y values is the **range**. We plot the x and y values in the $x - y$ **plane**. A plane is an infinitesimally flat surface that extends from $-\infty$ to ∞ in every direction.

Example

Graph: $4x - y = 2$

First isolate the dependent variable. Rewriting:
$$y = 4x - 2$$

Let's pick $x = 0$, and $x = 1$.
    When $x = 0, y = 4(0) - 2 = -2$.
    When $x = 1, y = 4(1) - 2 = 2$.

The graph is to the right.

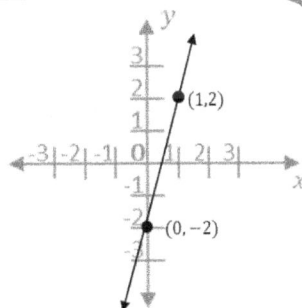

Graph of $4x - y = 2$.

Some Common Functions:

Sometimes you have to sketch a graph without a computer or graphing calculator. A good question is: How do you know how many points you need to plot to make sure you didn't miss anything critical? Well, being familiar with the general shape an equation would produce helps a lot. The following discussion outlines some common functions worth knowing.

Linear: When the highest order of an equation is 1, the equation is a **linear** expression. Consider the simple example of $y = ax$ where a is a constant. For every x and $-x$ pair, the answer for y is equal and opposite. You would expect a graph that decreases to the left and increases to the right from (0,0). The figure to the right shows the four variations of the simple linear expression. The next chapter will deal with these relations in more detail.

| $y = ax$ | $y = -ax$ |

Simple linear expressions.

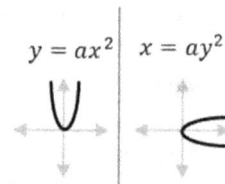

Quadratic: When the highest order of an equation is 2, the equation is a **quadratic** expression. Consider the simple example of $y = ax^2$ where a is a constant. For every x and $-x$ pair, the answer for y is the same. You would expect a graph that increases from (0,0) and is symmetric about the y-axis. Indeed this is true. The figure to the right shows the four variations of the simple quadratic. Chapter 7 will deal with these relations in more detail.

| $y = ax^2$ | $x = ay^2$ |

| $y = -ax^2$ | $x = -ay^2$ |

Simple quadratic expressions.

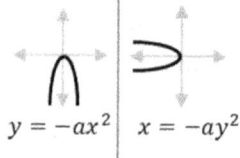

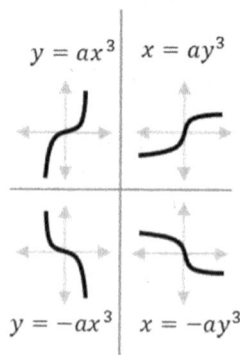

Cubic: When the highest order of an equation is 3, the equation is a **cubic** expression. Consider the simple example of $y = ax^3$ where a is a constant. For every x and $-x$ pair, the answer is equal and opposite. You would expect a graph that decreases to the left and increases to the right from (0,0). The figure to the right shows the four variations of the simple cubic.

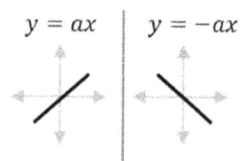

| $y = ax^3$ | $x = ay^3$ |

| $y = -ax^3$ | $x = -ay^3$ |

Simple cubic expressions.

Example

Graph: $$y = x^3 + 2x^2 + x$$

This is a cubic expression that combines the simple cubic, quadratic, and linear expressions we discussed thus far. We might expect the cubic to dominate with larger $|x|$ values since it increases (or decreases) fastest. The graph is shown to the right.

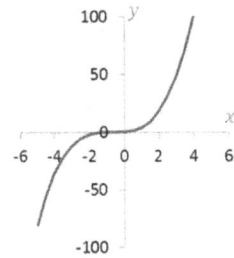

$y = x^3 + 2x^2 + x.$

Logarithms: Logarithms such as $y = \log_b ax$ are not defined for $x \leq 0$. The three points of interest are:
- For $0 < ax < 1$, the function increases rapidly from $-\infty$.
- At $ax = 1, y = 0$.
- For $ax > 1$, the function increases less rapidly.

As the base, b, gets smaller, the curve from $0 < ax < 1$ gets less steep, and rises higher for $ax > 1$. Two log plots are shown to the right.

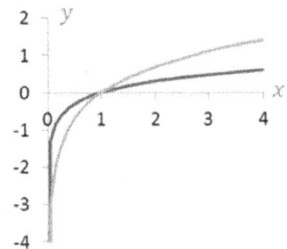

Plot of $y = \log x$ in dark grey and $y = \ln x$ in light grey.

Exponentials: Exponentials like $y = b^{ax}$ have three key regions:
- For $x > 1/a$, the function increases very fast.
- For $0 < x < 1/a$, the exponent is a fraction which takes roots of the number b. The function will vary between $b^0 = 1$ and $b^1 = b$.
- For $x < 0$, the negative exponent means a plot of $1/b^{|ax|}$ which decreases to almost $y = 0$.

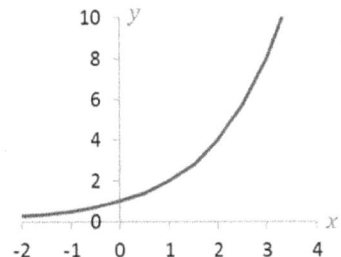

Plot of $y = 2^x$.

For exponentials like $y = b^{-x}$, the above pattern is reversed. This function is called a **decay** function because y decreases to nearly 0 as x increases. The special case of the natural exponential to a negative x power is found in many real life applications. The function $y = e^{-ax}$ is plotted to the right. The constant a is called the **attenuation constant** which determines how fast the function decays. On a graph, the attenuation constant can be found by locating where $y = \frac{1}{e} \approx 0.37$ and noting that the x coordinate must then equal $1/a$.

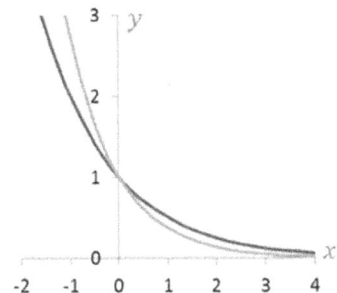

Plot of $y = 2^{-x}$ in dark grey and $y = e^{-x}$ in light grey.

31

Graph: $\qquad y = \log 2x$

The graph is shown to the right. Notice it rises rapidly from $y = -\infty$ to $y = 0$. The point where $y = 0$ is the point where $ax = 1$, or $x = 1/2$.

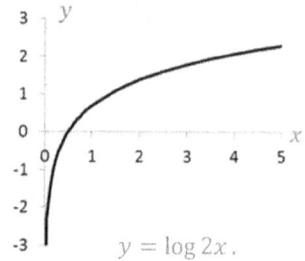

$y = \log 2x$.

Example

What is the attenuation constant for the graph of the natural exponential below:

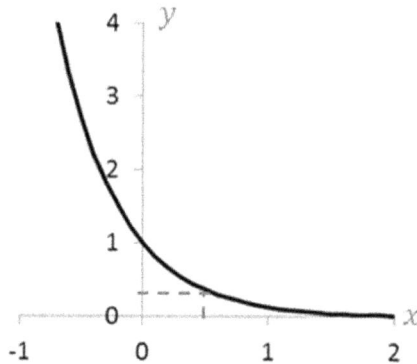

When $y \approx 0.37$, $x = 1/2$. The attenuation constant is 2, and the equation plotted is $y = e^{-2x}$.

Chapter 6: Lines

Distance, Midpoint, & Slope:

As you learned in Geometry, lines can be defined by at least two points. The two points can be written as (x_1, y_1) and (x_2, y_2). The y_1 and y_2 values are the results for y we get from the equation $y = f(x)$ after plugging in the x_1 and x_2 values.

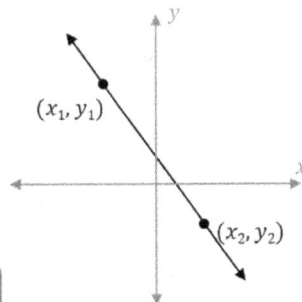

The **distance** between the two points is given by:

$$distance = d = \sqrt{(x_2 - x_1)^2 + (y_2 - y_1)^2}$$

A line defined by two points: (x_1, y_1) and (x_2, y_2).

The **midpoint** between the two points is given by:

$$midpoint = \left(\frac{x_1 + x_2}{2}, \frac{y_1 + y_2}{2}\right)$$

Example

Find the distance and midpoint between:
$(1, -2)$ and $(3, 4)$

Plugging these points into the formulas we get:

$$d = \sqrt{(x_2 - x_1)^2 + (y_2 - y_1)^2}$$

$$= \sqrt{(3 - 4)^2 + (-2 - 1)^2}$$

$$= \sqrt{(-1)^2 + (-3)^2} = \sqrt{1 + 9} = \sqrt{10}$$

Given points are in black, midpoint is in grey.

$$midpoint = \left(\frac{x_1 + x_2}{2}, \frac{y_1 + y_2}{2}\right)$$

$$= \left(\frac{1+3}{2}, \frac{-2+4}{2}\right) = \left(\frac{4}{2}, \frac{2}{2}\right) = (2,1)$$

Don't worry about choosing which is point 1 and which is point 2. As long as you are consistent (don't mix up the x's or y's), it doesn't matter.

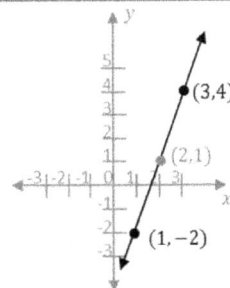

If we switch the labeling of the points, we get the same result:

$$d = \sqrt{(4 - 3)^2 + (1 - (-2))^2}$$

$$= \sqrt{(1)^2 + (3)^2} = \sqrt{10}$$

$$midpoint = \left(\frac{3+1}{2}, \frac{4+(-2)}{2}\right)$$

$$= \left(\frac{4}{2}, \frac{2}{2}\right) = (2,1)$$

When we talk about lines, we often want to know the **slope** of the line, i.e. the direction the line is heading if traveling from left to right. Slope is more formally described by the change in height (Δy) over the change in length (Δx). The slope between two points is given by:

$$slope = m = \frac{\Delta y}{\Delta x} = \frac{y_2 - y_1}{x_2 - x_1}$$

It does not matter which point you pick as (x_1, y_1) or (x_2, y_2) as long as you are consistent (don't mix the x's and y's between pairs). Also, you can pick any two points on a straight line to calculate the slope.

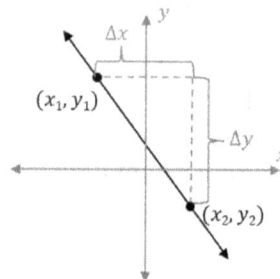

Calculating Slope.

Example

Find the slope between:
$$(1, -2) \text{ and } (3, 4)$$

Plugging these points into the formula we get:

$$m = \frac{\Delta y}{\Delta x} = \frac{y_2 - y_1}{x_2 - x_1} = \frac{-2 - 4}{1 - 3} = \frac{-6}{-2} = 3$$

If we switch the order of the points, we get the same result:

$$m = \frac{4 - (-2)}{3 - 1} = \frac{6}{2} = 3$$

A slope of +3 means that the line points upward (positive) and has a rise to run ratio of 3:1 (goes up a distance of 3 for every 1 in length).

Equations for Lines:

What is nice about slope is that it allows us to get the equation for a line. Let's imagine that we know the slope of the line, m, but we only know one point (a, b). Let's plug these into the slope equation:

$$m = \frac{y_2 - y_1}{x_2 - x_1} = \frac{y - b}{x - a}$$

Notice that we dropped the subscripts for the point (x_2, y_2) above since we no longer need them.

If we now solve for $y - b$, we get:

$$y - b = m(x - a)$$

This is the **Point-Slope Form** for the equation of a line because it contains a point and a slope.

Example

Find the equation for the line between:
$$(1, -2) \text{ and } (3,4)$$

We already know the slope is 3 from the prior example. Pick one of the points above for (a, b), and plug it into the point-slope form:
$$y + 2 = 3(x - 1)$$

If we were to plug in $(x, y) = (3,4)$, the equation should be true:
$$4 + 2 = 3(3 - 1)$$
$$6 = 6$$

The point-slope form of an equation of a line can give us any point on the line.

Equivalently, we could have chosen $(a, b) = (3,4)$:
$$y - 4 = 3(x - 3)$$

If we plug in $(x, y) = (1, -2)$, we get:
$$-2 - 4 = 3(1 - 3)$$
$$-6 = -6$$
which is true.

Now consider the special case of choosing (a, b) as the point where the line crosses the y-axis. This forces $a = 0$, so the point is now $(0, b)$. In this case, b has a special name and is called the **y-intercept** of the line.

Plugging this point into the point-slope form of the equation for a line gives:
$$y - b = m(x - 0)$$

Solving for y gives:

$$y = mx + b$$

This is the **Slope-Intercept Form** for the equation of a line as it contains the slope and the y-intercept.

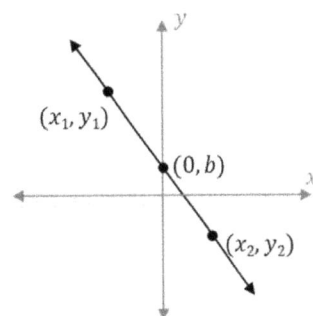

The y-intercept of a line is the point where the line crosses the y-axis, i.e. where $x = 0$.

Example

Find the slope-intercept equation for the line between:
$$(1, -2) \text{ and } (3,4)$$

From the prior example we have:
$$y + 2 = 3(x - 1)$$

If we solve this equation for y, we get:
$$y + 2 = 3x - 3$$
$$y = 3x - 5$$

So, the y-intercept occurs at (0,-5).

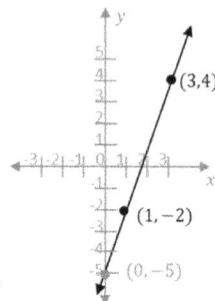

The y-intercept is in grey.

Note: The equation
$$y - 4 = 3(x - 3)$$
would give the same result.

Example

Plot the line given by:
$$2y - 4x = -4$$

First we put the equation in slope-intercept form:
$$2y = 4x - 4$$
$$y = 2x - 2$$

This equation tells us that the slope is 2, and the y-intercept is -2. We plot the point $(0, -2)$, then go up two and over one $(m = +2)$ for a second point. The line connects the dots.

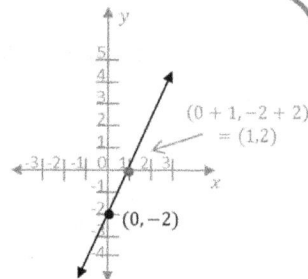

$(0 + 1, -2 + 2)$
$= (1,2)$

$(0, -2)$

The y-intercept is in black, the second point determined from the slope is in grey.

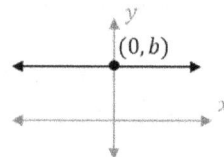

For a **horizontal line** (back and forth), all y's are equal, so $\Delta y = y_2 - y_1 = 0$ which forces $m = {}^{\Delta y}/_{\Delta x} = 0$. The equation for the line becomes

$$y = b.$$

$(0, b)$

Horizontal line.

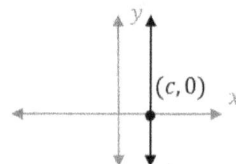

For a **vertical line** (up and down), all x's are equal, so $\Delta x = x_2 - x_1 = 0$ which forces $m = {}^{\Delta y}/_{\Delta x} = \infty$. There is no point-slope or slope-intercept form for the equation of a vertical line. The equation is simply

$$x = c.$$

$(c, 0)$

Vertical line.

Finally, we consider parallel and perpendicular lines.
Parallel lines lie in the same plane and never intersect.
Perpendicular lines lie in the same plane and intersect at a right angle.

> **Two lines are parallel iff $m_1 = m_2$.**

In other words, if the slope of line 1 is equal to the slope of line 2, then the two lines are parallel.

> **Two lines are perpendicular iff $m_1 = -1/m_2$.**

In other words, if the slope of line 1 is the negative inverse of that of line 2, then the two lines are perpendicular.

Example

Find the equation of a line that is parallel to $y = 2x + 3$ and goes through point $(1,4)$.

We are given a line in slope-intercept form. To be parallel, then, our line must have $m = 2$. We have one point to work from, so we can put our equation in point-slope form then solve for the slope-intercept form:

$$y - 4 = 2(x - 1) = 2x - 2$$

$y = 2x + 2$ the y=intercept is $(0,2)$.

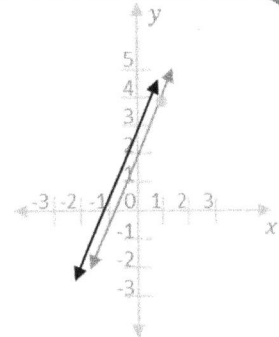

Original line is in black; parallel line is in grey.

Example

Find the equation of a line that is perpendicular to $y = 2x + 3$ and goes through point $(1,4)$.

Now we need $m = -1/2$. In point-slope and slope-intercept forms:

$$y - 4 = -\frac{1}{2}(x - 1) = -\frac{x}{2} + \frac{1}{2}$$

$y = -\frac{x}{2} + \frac{9}{2}$ the y=intercept is $\left(0, \frac{9}{2}\right)$.

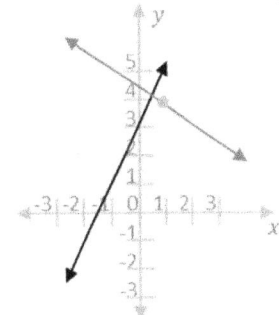

Original line is in black; perpendicular line is in grey.

Chapter 7: Circles & Conics

Circles:

A **circle** is a collection of planar points that are equidistant from a single point. The point is called the **center**, and the distance from the center to the circle edge is the **radius**. The **diameter** spans the circle and is twice the radius.

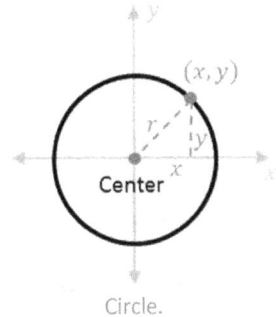

Circle.

Using our distance formula and the figure to the right, we see that any point on the circle (x, y), is a radius away from center. If the center is chosen to be at the origin at $(0,0)$, Pythagorean's Theorem gives us:

$$distance = r = \sqrt{x^2 + y^2}$$

Rewriting terms gives us the general equation for a circle centered at the origin:

$$x^2 + y^2 = r^2$$

What happens if we move the circle so that it is centered at (h, k)? In this case, we plug in $(x_1, y_1) = (x, y)$ and $(x_2, y_2) = (h, k)$ to get:

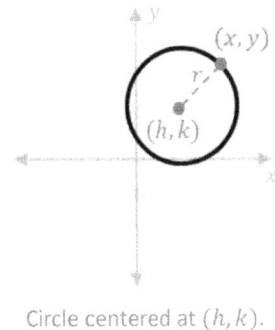

Circle centered at (h, k).

$$distance = r = \sqrt{(x - h)^2 + (y - k)^2}$$

Rewriting gives us the general equation for a circle of radius r centered at (h, k):

$$(x - h)^2 + (y - k)^2 = r^2$$

Example

Find the equation for a circle that has a diameter of 2 and is centered 3 units down from the origin.

The radius is $2/2 = 1$, and $h = 0, k = -3$.

Plugging in: $x^2 + (y + 3)^2 = 1$ which is plotted to the right.

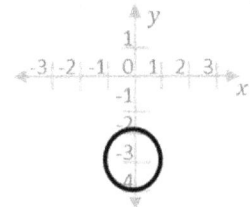

Circle for example.

Parabolas:

A **parabola** is a collection of planar points that are equidistant from a single point and a line. The point is called the **focus**, and the line is called the **directrix**. The parabola is shaped like an infinitely long U that opens away from the directrix toward the focus. Where the slope of the parabola changes direction is called the **vertex**.

Consider a parabola that faces straight upward and has a vertex at the origin. All of its points are equidistant from a point on the y-axis, $(0, p)$, and a line p away from the x-axis, $y = -p$. The distance from the point is given by:

$$distance\ to\ point = distance\ to\ line$$

$$\sqrt{(x-0)^2+(y-p)^2} = \sqrt{(x-x)^2+(y+p)^2}$$

$$x^2 + y^2 - 2py + p^2 = y^2 + 2py + p^2$$

$$x^2 = 4py; \ focus\ (0, p), directrix\ y = -p$$

The above is the equation for a parabola with focus $(0, p)$ and directrix $y = -p$. Similarly, the parabola that opens along the x-axis with a focus $(p, 0)$ and directrix $x = -p$ is given by:

$$y^2 = 4px; \ focus\ (p, 0), directrix\ x = -p$$

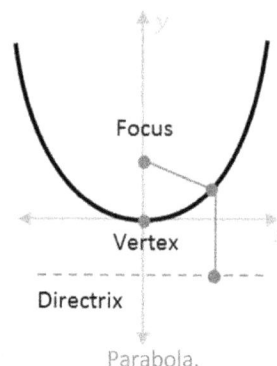

Focus
Vertex
Directrix

Parabola.

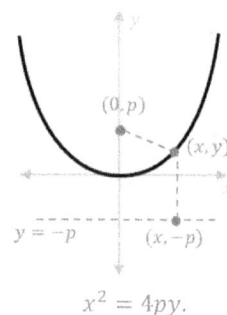

$(0, p)$
(x, y)
$y = -p$
$(x, -p)$

$x^2 = 4py.$

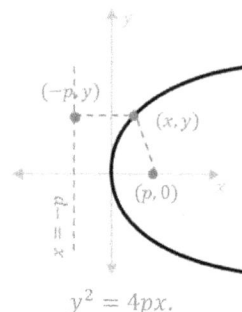

$(-p, y)$
(x, y)
$(p, 0)$
$x = -p$

$y^2 = 4px.$

| **Example** | Find the general equation of a parabola that opens toward the $-x$ direction. |

In this case, the focus and directrix switch positions with respect to the y-axis. The focus is now $(-p, 0)$, the directrix is $x = +p$, and the equation is $y^2 = -4px$.

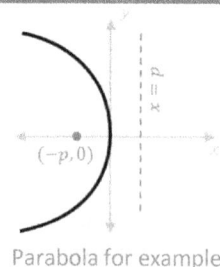

$(-p, 0)$
$x = p$

Parabola for example.

Ellipses:

An **ellipse** is a collection of planar points for which the sum of their distances to two fixed points is constant. In other words, if you tack a loose string to two points and pull the slack with a pencil, an oval will be traced out. The points are called the **foci**, and the midpoint between them is the **center** of the ellipse (the origin for the ellipse to the right).

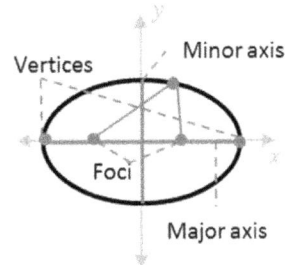

Ellipse with center at (0,0).

If you draw a line from edge to edge and through the foci, you trace out the largest dimension of the ellipse. This is called the **major axis**, and its endpoints on the ellipse are called **vertices**. If you draw a line from edge to edge through the center and perpendicular to the major axis, you find the **minor axis**.

Let's look at a simple ellipse that is centered at the origin with foci located along the x-axis at points $(-c, 0)$ and $(c, 0)$. If the sum of lengths is set equal to $2a$, where $a > c$, we have:

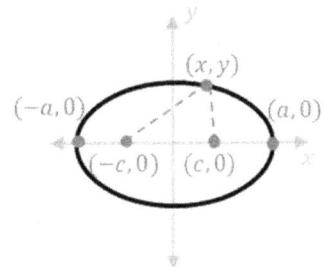

Ellipse with foci at $(\pm c, 0)$.

$$\text{distance to point 1} + \text{distance to point 2} = 2a$$

$$\sqrt{(x+c)^2+(y-0)^2}+\sqrt{(x-c)^2+(y-0)^2}= 2a$$

$$\sqrt{(x+c)^2+(y)^2}= 2a - \sqrt{(x-c)^2+(y)^2}$$

Rearrange.

$$\begin{aligned} x^2 + 2xc + c^2 + y^2 \\ = 4a^2 - 4a\sqrt{(x-c)^2 + (y)^2} + x^2 - 2xc + c^2 + y^2 \end{aligned}$$

Square both sides.

$$a\sqrt{(x-c)^2 + (y)^2} = a^2 - xc$$

Simplify.

$$a^2(x^2 - 2xc + c^2) + a^2y^2 = a^4 - 2a^2xc + x^2c^2$$

Square both sides.

$$x^2(a^2 - c^2) + a^2y^2 = a^2(a^2 - c^2)$$

Simplify & group.

$$\frac{x^2}{a^2} + \frac{y^2}{a^2-c^2} = 1$$

Rearrange.

Now we define a length, b, such that $b^2 = a^2 - c^2$. Finally, we have the general equation for an ellipse with foci at $(\pm c, 0)$ and sum of lengths equal to $2a$:

$$\frac{x^2}{a^2} + \frac{y^2}{b^2} = 1; \;\; foci \; at \; (\pm c, 0)$$

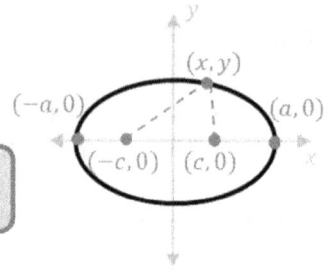

Ellipse with foci at $(\pm c, 0)$.

The endpoints of the major axis lie on $(\pm a, 0)$, and the endpoints of the minor axis lie on $(0, \pm b)$.

Using the same reasoning, an ellipse with foci at $(0, \pm c)$ and sum of lengths equal to $2a$ is given by:

$$\frac{x^2}{b^2} + \frac{y^2}{a^2} = 1; \;\; foci \; at \; (0, \pm c)$$

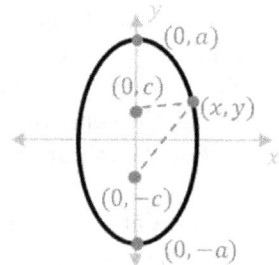

Ellipse with foci at $(0, \pm c)$.

The endpoints of the major axis lie on $(0, \pm a)$, and the endpoints of the minor axis lie on $(\pm b, 0)$.

As a final note, an ellipse will be almost circular if $a \gg c$, and nearly flat if a is only slightly larger than c. The **eccentricity**, E, of an ellipse is the ratio of c to a given by:

$$E = \frac{c}{a} = \frac{\sqrt{a^2 - b^2}}{a}$$

Since $a > c, 0 < E < 1$. The larger the eccentricity, the flatter the ellipse.

Example

Sketch the ellipse whose axes lie between $(0, \pm 3)$ and $(\pm 1, 0)$. State the equation.

Since $3 > 1$, the major, or longer, axis lies along the y direction. Then $a = 3, b = 1$, and the equation is:

$$\frac{x^2}{b^2} + \frac{y^2}{a^2} = 1$$

$$x^2 + \frac{y^2}{9} = 1$$

The plot is to the right.

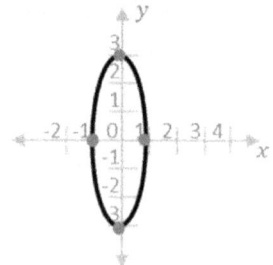

Ellipse for example.

Hyperbolas:

A **hyperbola** is a collection of planar points for which the difference of their distances to two fixed points is constant.. The points are called the **foci**, and the midpoint between them is the **center** (the origin for the hyperbola shown).

A hyperbola looks like two parabolas facing away from each other. The points where the curves change direction are called the **vertices**. The **transverse axis** is the line that would connect the foci. The **conjugate axis** is the line midway between vertices and perpendicular to the transverse axis.

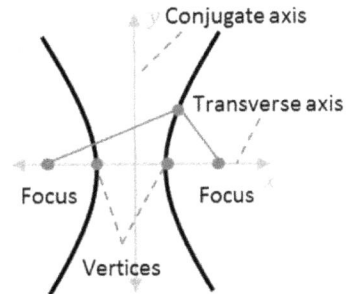

Hyperbola with center at (0,0).

We start with a simple hyperbola with the y-axis as the transverse axis, and foci on the x-axis at points $(\pm c, 0)$. If the difference in lengths is set equal to $2a$, where $a < c$, the vertices will lie at $(\pm a, 0)$. Using the same logic as for the ellipse (we skip the math here), we have:

$$|distance\ to\ point\ 1 - distance\ to\ point\ 2| = 2a$$

$$\left|\sqrt{(x+c)^2+(y-0)^2} - \sqrt{(x-c)^2+(y-0)^2}\right| = 2a$$

⋮ Skipping the math.

$$\frac{x^2}{a^2} - \frac{y^2}{c^2-a^2} = 1$$

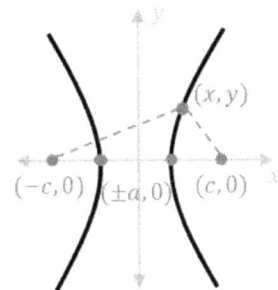

Hyperbola with foci at $(\pm c, 0)$.

We define b this time so that $b^2 = c^2 - a^2$ and get the general equation for a hyperbola with foci at $(\pm c, 0)$ and vertices at $(\pm a, 0)$:

$$\frac{x^2}{a^2} - \frac{y^2}{b^2} = 1; \quad foci\ at\ (\pm c, 0)$$

A hyperbola with foci at $(0, \pm c)$ and vertices at $(0, \pm a)$ is:

$$\frac{y^2}{a^2} - \frac{x^2}{b^2} = 1; \quad foci\ at\ (0, \pm c)$$

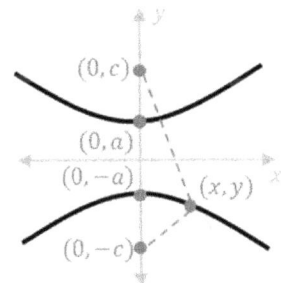

Hyperbola with foci at $(0, \pm c)$.

To sketch a hyperbola, knowing the asymptotes is very helpful. The **asymptotes** are lines that the hyperbola **branches** (or curves) approach but never intersect. They are the closest lines you can draw that will never touch the hyperbola. For a hyperbola opening on the x-axis, the asymptote lines are $y = \pm\frac{b}{a}x$. For a hyperbola opening on the y-axis, the asymptote lines are $y = \pm\frac{a}{b}x$.

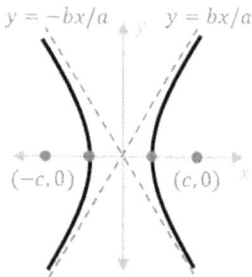

Hyperbola opening on x-axis.

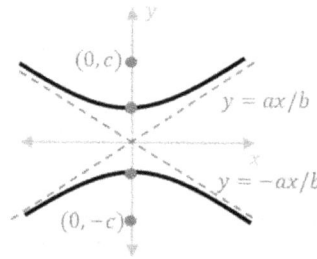

Hyperbola opening on y-axis.

 **Example**

Sketch the equation $\frac{y^2}{4} - x^2 = 1$.

The y denominator is bigger, so $a = 2, b = 1$. The asymptotes are:

$$y = \pm\frac{a}{b}x = \pm2x.$$

The vertices lie at $(0, \pm2)$.

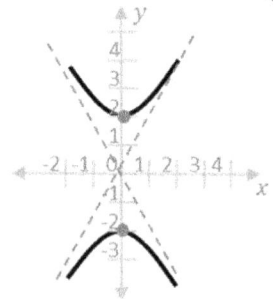

Hyperbola for example.

A final note on these general shapes: parabolas, ellipses, and hyperbolas are called **conic sections**, or **conics** for short. The reason is that they can be obtained by an intersecting plane at different angles relative to a cone as shown to the right.

Parabola.

Ellipse.

Hyperbola.

Circle & Conic Summary:

Equation	Parameters	Points of Interest	Sketch
$(x-h)^2+(y-k)^2=r^2$	$r>0$	Radius $=r$ Center $=(h,k)$	⭕
$x^2=4py$	$-\infty<p<\infty$	Focus $=(0,p)$ Directrix: $y=-p$	
$y^2=4px$		Focus $=(p,0)$ Directrix: $x=-p$	
$\dfrac{x^2}{a^2}+\dfrac{y^2}{b^2}=1$	$a>b>0$; $a>c>0$ $b^2=a^2-c^2$	Maj Ax $=(\pm a,0)$ Min Ax $=(0,\pm b)$ Focus $=(\pm c,0)$	
$\dfrac{y^2}{a^2}+\dfrac{x^2}{b^2}=1$		Maj Ax $=(0,\pm a)$ Min Ax $=(\pm b,0)$ Focus $=(0,\pm c)$	
$\dfrac{x^2}{a^2}-\dfrac{y^2}{b^2}=1$	$c>a>b>0$ $b^2=c^2-a^2$	Vertices $=(\pm a,0)$ Focus $=(\pm c,0)$ Asymptote: $y=\pm bx/a$	
$\dfrac{y^2}{a^2}-\dfrac{x^2}{b^2}=1$		Vertices $=(0,\pm a)$ Focus $=(0,\pm c)$ Asymptote: $y=\pm ax/b$	

Chapter 8: Systems of Equations

Systems of Independent Equations:

An interesting problem arises when we are given two or more equations to solve at once, i.e. a **system of equations**. We can come up with solutions for each equation, but what we really want to know is what solutions are common to all of the equations. This is referred to as **simultaneously solving** the system of equations.

For a two equation system, if S_1 is the solution set for one equation, and S_2 is the solution set for the second equation, then we want to find:

$$S_1 \cap S_2$$

There are two things we need to check before even attempting to do this:
1. We have as many variables as we have equations (people often say, "as many equations as unknowns").
2. All of the equations are **independent** of each other.

The first criterion is easy enough to check; if you have two equations, you better have two variables. The second criterion calls for the equations to be independent. This means that you can't derive one equation from another.

When you solve single equations, you perform operations to both sides of the equal sign. If you can do this to one equation to get the other equation, then the two equations are **dependent** on one another. For example:

$$y = 2x + 4 \text{ and } 2y = 4x + 8$$

are dependent equations because you can multiply the first equation by 2 to get the second equation.

However, if you have successfully met the criteria above, then you can proceed to simultaneously solve the system using any of three following techniques:

Combination, **Substitution**, or **Graphing**

Combination:

The first method to simultaneously solve a system of equations is to combine them through addition or subtraction to get rid of one variable. This gives a third equation with one unknown which can then be solved. Plugging that solution into either of the original two equations will give the values for the other variable.

Example

Solve the system of equations:
$$y = 4x + 3; \quad 2y = 4x - 2$$

Let's try to get rid of the y variable by subtracting 2 times the first equation from the second equation:

$$\begin{array}{r} 2y = 4x - 2 \\ - 2 \times [y = 4x + 3] \end{array} \quad \Rightarrow \quad \begin{array}{r} 2y = 4x - 2 \\ + \quad - 2y = -8x - 6 \\ \hline 0 = -4x - 8 \end{array}$$

Solving $0 = -4x - 8$ gives us $x = -2$. Plugging that result back into the first equation gives:
$$y = 4(-2) + 3 = -5.$$

The solution to both equations is $(x, y) = (-2, -5)$.

Or, you can also plug $x = -2$ into the 2nd equation:
$$2y = 4(-2) - 2 = -10$$
$$y = {-10}/{2} = -5$$

Example

Solve the system of equations:
$$3y = x + 2; \quad y = x - 1$$

Get rid of the x variable by subtracting the first equation from the second:

$$\begin{array}{r} 3y = x + 2 \\ - \quad [y = x - 1] \\ \hline 2y = \quad 3 \end{array}$$

Solving $2y = 3$ gives us $y = \frac{3}{2}$. Plugging this result into the second equation gives:
$$\frac{3}{2} = x - 1 \rightarrow x = 1 + \frac{3}{2} = \frac{5}{2}$$

The solution to both equations is $(x, y) = (\frac{5}{2}, \frac{3}{2})$.

Substitution:

The second method to simultaneously solve a system of equations is to use substitution:
- Isolate one variable in the first equation so that you get an expression for that variable ($y = \cdots$).
- Substitute the expression found in step one into all occurrences of the variable in the second equation.
- You now have one equation with one unknown. Solve this equation to find the values for this variable.
- Plug the result back into the first equation to find the values for the first variable.

Example

Solve the system of equations:
$$y = 4x + 3; \quad 2y = 4x - 2$$

The first equation is already solved in terms of one variable: $\qquad y = 4x + 3$

So, we can plug in $(4x + 3)$ for y in the second equation and solve:
$$2y = 4x - 2$$
$$2(4x + 3) = 4x - 2$$
$$8x + 6 = 4x - 2$$
$$x = -2$$

Plug this into either of the first equations and get:
$$y = 4(-2) + 3 = -5$$

Example

Solve the system of equations:
$$3y = x + 2; \quad y = x - 1$$

Plug in the second equation into the first:
$$3(x - 1) = x + 2$$
$$2x = 5 \;\rightarrow\; x = \frac{5}{2}$$

Plugging back into the second equation:
$$y = \frac{5}{2} - 1 = \frac{3}{2}$$

Graphing Solutions:

The third method to simultaneously solve a system of equations is to graph both equations and literally look for the intersections. The drawback is that the solution is only as accurate as your drawing.

Example

Solve the system of equations:
$$y = 4x + 3; \quad 2y = 4x - 2$$

The first equation is in slope-intercept form:
$$y = 4x + 3$$

We plot the point $(0,3)$, then use the slope, $m = 4$, to plot the line.

The second equation can be modified to fit into slope-intercept form:
$$2y = 4x - 2$$
$$y = 2x - 1$$

We plot the point $(0, -1)$, then use the slope, $m = 2$, to plot the line.

The plot for both equations is to the right. By inspection, we can see that they overlap at the point $(x, y) = (-2, -5)$.

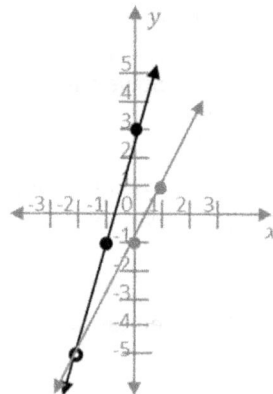

The first line is in black, the second in grey. The intersection is circled.

Example

Solve the system of equations:
$$3y = x + 2; \quad y = x - 1$$

Putting both equations into slope-intercept form:
$$y = \frac{x}{3} + \frac{2}{3};$$

$$y = x - 1$$

Graphing the equations gives the intercept at:

$$(x, y) = \left(\frac{5}{2}, \frac{3}{2}\right)$$

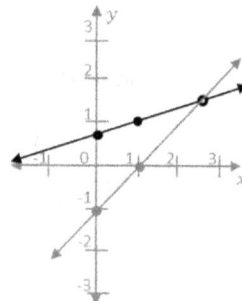

The first line is in black, the second in grey. The intersection is circled.

Chapter 9: Congruent Transformations

Translations:

You may remember from Geometry that there are three types of **congruent transformations**, moving a shape while retaining its original form:
- **translations** (sliding the object)
- **reflections** (flipping the object)
- **rotations** (spinning the object).

Let's say we have a function $y = f(x)$ that we want to move h units right and k units up. The formula to do so is simply:

> **To move $y = f(x)$ h units right and k units up, plot**
>
> $$y - k = f(x - h).$$

This means that everywhere you see y, plug in $(y - k)$, and everywhere you see x, plug in $(x - h)$.

Example

Move the line $y = 4x + 2$
    2 units to the right and 4 units up.

We plug in $x - 2$ and $y - 4$ to get:

$$y - 4 = 4(x - 2) + 2$$

$$y = 4x - 2$$

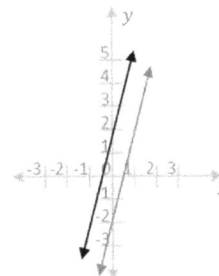

Translated line is in grey.

The above formula works for non-functions, as well. Recall from Chapter 7 that a circle centered at the origin is:

$$x^2 + y^2 = r^2.$$

But, a circle centered at (h, k) is:

$$(x - h)^2 + (y - k)^2 = r^2.$$

Using the translation formula, we can move parabolas, ellipses, and hyperbolas, as well. In order to find our new foci, vertices, etc., we use the translation formula again as outlined in the insert.

For parabolas:

$$(x - h)^2 = 4p(y - k)$$

focus $(h, k + p)$, **directrix** $y = k - p$, **vertex** (h, k).

$$(y - k)^2 = 4p(x - h)$$

focus $(h + p, k)$, **directrix** $x = h - p$, **vertex** (h, k).

For ellipses $(b^2 = a^2 - c^2)$:

$$\frac{(x-h)^2}{a^2} + \frac{(y-k)^2}{b^2} = 1$$

foci $(h \pm c, k)$, **axes at** $(h \pm a, k)$, $(h, k \pm b)$.

$$\frac{(x-h)^2}{b^2} + \frac{(y-k)^2}{a^2} = 1$$

foci at $(h, k \pm c)$, **axes at** $(h, k \pm a)$, $(h \pm b, k)$.

For hyperbolas $(b^2 = c^2 - a^2)$:

$$\frac{(x-h)^2}{a^2} - \frac{(y-k)^2}{b^2} = 1$$

foci $(h \pm c, k)$, **vertices** $(h \pm a, k)$,
asymptote $y = \pm\frac{b}{a}(x - h) + k$.

$$\frac{(y-k)^2}{a^2} - \frac{(x-h)^2}{b^2} = 1$$

foci $(h, k \pm c)$, **vertices** $(h, k \pm a)$,
asymptote $y = \pm\frac{a}{b}(x - h) + k$.

NOTE:

To find the foci, vertices, etc., we simply plug in $x - h$ and $y - k$ into these expressions, as well. For example, a focus at $(p, 0)$ is a focus where $x = p, y = 0$.

Plugging in our new expressions:
$$x - h = p \rightarrow x = h + p$$
$$y - k = 0 \rightarrow y = k$$
Focus is now $(h + p, k)$.

So, these formulas are handy, but if you know the originals, you can derive these easily.

> **Example**
>
> Sketch the equation:
> $$9x^2 - 18x + 25y^2 - 100y = 116$$
>
> This may seem unrecognizable, but looking at our other forms, we know circles and conics have x^2 and y^2, so that's likely what this is. Let's start by separating the coefficients from x^2 and y^2.
>
> $$9(x^2 - 2x) + 25(y^2 - 4y) = 116$$
>
> In most of our equations, we have quantities that look like $(x - h)^2$ and $(y - k)^2$. We can add quantities to both sides of our equations to get something factorable:
>
> $$9(x^2 - 2x + 1) + 25(y^2 - 4y + 4)$$
> $$= 116 + 9 + 100$$
>
> $$9(x - 1)^2 + 25(y - 2)^2 = 225$$
>
> For this to be a circle, the coefficient in front of $(x - h)^2$ and $(y - k)^2$ must be the same. Since this is not the case and there is a plus sign between the two expressions, we suspect we have an ellipse. We now divide by 225 to get a 1 on the right side:
>
> $$\frac{(x-1)^2}{25} + \frac{(y-2)^2}{9} = 1$$
>
> Indeed, we have an ellipse with:
> $$a = 5, b = 3, h = 1, \text{ and } k = 2.$$
>
> The major axes are at:
> $$(h \pm a, k) \text{ or } (-4,2) \text{ and } (6,2).$$
>
> The minor axes are at:
> $$(h, k \pm b) \text{ or } (1, -1) \text{ and } (1,5).$$

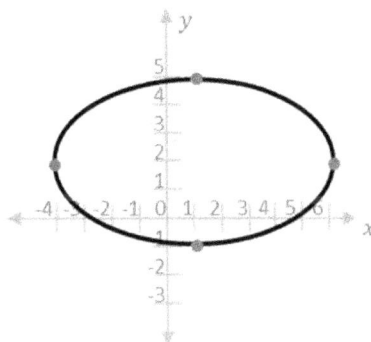

Plot for the example.

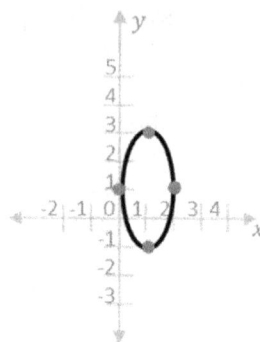

Find the equation for the figure to the right.

Example

We have an ellipse with major axes at $(1,3)$ and $(1,-1)$. We see from the orientation that the major axes endpoints are at $(h, k \pm a)$, so $h = 1$. Writing the equations for $k \pm a$ and adding them together:

$$k + a = 3; \ k - a = -1$$
$$2k = 2 \rightarrow k = 1 \rightarrow a = 2$$

The minor axes are at $(0,1)$ and $(2,1)$. at $(h \pm b, k)$. We know $h = 1$ from above, so $b = 1$.

Figure for example.

Our equation is:

$$\frac{(x-h)^2}{b^2} + \frac{(y-k)^2}{a^2} = 1$$

$$\frac{(x-1)^2}{1} + \frac{(y-1)^2}{4} = 1$$

Reflections About the Axes:

Reflections about the axes are straightforward:

> **To reflect $f(x)$ about the y-axis $(x = 0)$:**
> **Plot** $\ \ y = f(-x)$;
>
> **To reflect $f(x)$ about the x-axis $(y = 0)$:**
> **Plot** $\ \ -y = f(x)$.

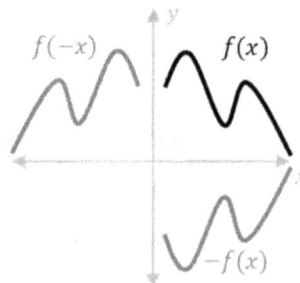

Reflecting about the axes.

In other words, to reflect about the y-axis, plug in $-x$ for x. To reflect about the x-axis, plug in $-y$ for y. The above relation holds for non-functions, as well.

Example

Reflect $\frac{(x-1)^2}{1} + \frac{(y-1)^2}{4} = 1$ about the y-axis and the x-axis.

Reflection about the y-axis:
We plug in $-x$ for x to get:

$$\frac{(-x-1)^2}{1} + \frac{(y-1)^2}{4} = 1$$

Simplifying our x terms:

$$(-x-1)^2 = x^2 + 2x + 1 = (x+1)^2$$

$$\frac{(x+1)^2}{1} + \frac{(y-1)^2}{4} = 1$$

We have an ellipse with $h = -1, k = 1, a = 2,$ and $b = 1$. The major axes are along the y-axis (y term denominator is largest) at $(h, k \pm a)$: $(-1, -1)$, $(-1, 3)$. The minor axes are at $(h \pm b, k)$: $(0, 1)$ and $(-2, 1)$. The figure is plotted to the right.

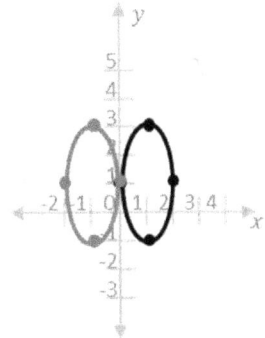

Reflection about y-axis:
Reflected object is in grey.

Reflection about the x-axis:
We plug in $-y$ for y to get:

$$\frac{(x-1)^2}{1} + \frac{(-y-1)^2}{4} = 1$$

$$\frac{(x-1)^2}{1} + \frac{(y+1)^2}{4} = 1$$

We have an ellipse with $h = 1, k = -1, a = 2,$ and $b = 1$. The major axes are at $(h, k \pm a)$: $(1, -3)$, $(1, 1)$. The minor axes are at $(h \pm b, k)$: $(0, -1)$ and $(2, -1)$. The figure is plotted to the right.

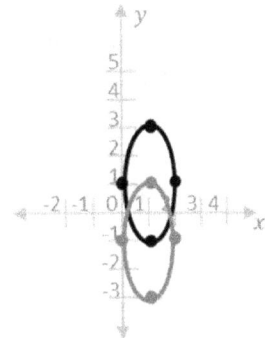

Reflection about the x-axis:
Reflected object is in grey.

Rotations:

Because most Algebra 2 students have had some access to trigonometry at this stage of the course, rotations are usually covered now. If you have not had any trigonometry, yet, skip to Chapter 10, and then come back to this section.

Rotations are straightforward, but they sometimes require a lot of bookkeeping. That's because the substitution formulas for equations and vertices can be quite lengthy.

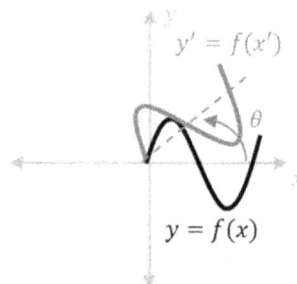

Counterclockwise rotation.

> **To rotate $f(x)$ $\theta°$ counterclockwise from the x-axis, plot $y' = f(x')$ where:**
>
> $$x' = x\cos\theta + y\sin\theta\,;$$
> $$y' = -x\sin\theta + y\cos\theta\,.$$

The following table can be quite helpful (and easy to remember) with common rotations. The complete table is in Chapter 10.

	0°	30°	45°	60°	90°
$\sin\theta$	$\dfrac{\sqrt{0}}{2}$	$\dfrac{\sqrt{1}}{2}$	$\dfrac{\sqrt{2}}{2}$	$\dfrac{\sqrt{3}}{2}$	$\dfrac{\sqrt{4}}{2}$
$\cos\theta$	$\dfrac{\sqrt{4}}{2}$	$\dfrac{\sqrt{3}}{2}$	$\dfrac{\sqrt{2}}{2}$	$\dfrac{\sqrt{1}}{2}$	$\dfrac{\sqrt{0}}{2}$

Example

Rotate $\dfrac{(x-1)^2}{1} + \dfrac{(y-1)^2}{4} = 1$ counterclockwise 90°.

Since $\sin 90° = 1, \cos 90° = 0$, we have $x' = y$; $y' = -x$. Plugging in gives:

$$\frac{(y-1)^2}{1} + \frac{(-x-1)^2}{4} = 1 \rightarrow \frac{(y-1)^2}{1} + \frac{(x+1)^2}{4} = 1$$

Now the major axis is in the x direction (larger denominator). We plot an ellipse with $h = -1$, $k = 1, b = 1$, and $a = 2$. The axes are located at $(-3,1), (1,1), (-1,2)$ and $(-1,0)$.

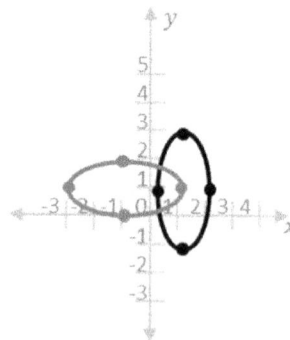

Rotation of 90°:
Rotated object is in grey.

Example Rotate $\frac{x^2}{1} + \frac{y^2}{4} = 1$ counterclockwise 45° and sketch the figure.

Since $\sin 45° = \cos 45° = \sqrt{2}/2$, then

$x' = \frac{\sqrt{2}}{2}(x + y)$; $y' = \frac{\sqrt{2}}{2}(y - x)$. Plugging in gives:

$$\frac{2(x+y)^2}{4} + \frac{2(y-x)^2}{16} = \frac{(x+y)^2}{2} + \frac{(y-x)^2}{8} = 1$$

The above equation is not in standard form since the x's and y's are mixed. For the original ellipse, the major axis was located at $(x = 0, y = \pm2)$. To find our new axes, we plug in our new x' and y' values (we'll consider point (0,2) first):

$$x' = \frac{\sqrt{2}}{2}(x + y) = 0; \quad y' = \frac{\sqrt{2}}{2}(y - x) = 2$$

$$y = -x; \quad y - x = 2\sqrt{2}$$

We have a system of 2 equations with 2 unknowns. Substituting $y = -x$ into the second equation:

$$x = -\sqrt{2} \rightarrow y = -x = \sqrt{2} \rightarrow (-\sqrt{2}, \sqrt{2}).$$

Doing the same for point $(0, -2)$ gives the point $(\sqrt{2}, -\sqrt{2})$. For the minor axis, we had points $(x = \pm1, y = 0)$. Point $(1,0)$ becomes:

$$x' = \frac{\sqrt{2}}{2}(x + y) = 1; \quad y' = \frac{\sqrt{2}}{2}(y - x) = 0$$

$$y + x = \sqrt{2}; \quad y = x \rightarrow \left(\frac{\sqrt{2}}{2}, \frac{\sqrt{2}}{2}\right)$$

Finally, point $(-1,0)$ becomes $\left(-\frac{\sqrt{2}}{2}, -\frac{\sqrt{2}}{2}\right)$. The figure is sketched to the right.

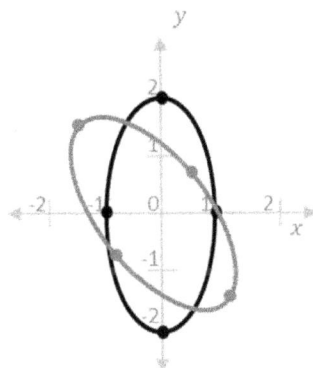

Rotation of 45°:
Rotated object is in grey.

Part 3: Trigonometry

Chapter 10: Trigonometric Functions

Sine, Cosine, & Tangent:

Triangles must obey some rigid properties that allow three sides to form a closed figure. **Right triangles** have the added property that one angle is 90° and the other two angles are **complimentary** (add to 90°).

In the figure to the right, we see a right triangle with one acute angle labeled. The angle is often referred to with the Greek symbol theta, θ. Once the angle is chosen, the sides can be labeled:
- the **hypotenuse** (HYP) across from the right angle,
- the side next to, or **adjacent** to, θ (ADJ), and
- the side **opposite** θ (OPP).

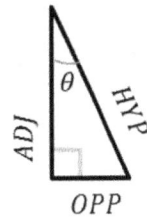

General right triangle.

Imagine keeping the ADJ side the same while increasing the angle θ. In order to keep a closed right triangle, the OPP and HYP sides must grow larger, and the third angle must get smaller.

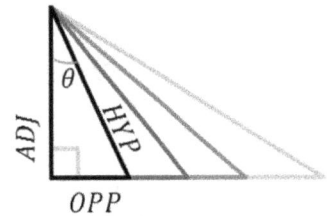

Some mathematician long ago discovered that these changes are very predictable. It doesn't matter what size the triangle is, as θ gets larger while holding ADJ fixed, the ratio of OPP to HYP follows the same pattern:

Right triangle with expanding angle θ.

- When $\theta = \sim 0°$,   OPP ≈ 0,   so OPP/HYP ≈ 0.
- When $\theta = \sim 90°$,   OPP $\approx$ HYP,   so OPP/HYP ≈ 1.

We go from the extreme of a vertical line to the extreme of a horizontal line in the two scenarios above. Solving for every value of θ in between, we will get a predictable outcome.

Let's call the ratio of OPP/HYP the **sine** of the angle, denoted $\sin \theta$. If we plot $\sin \theta$ against θ for $0° < \theta < 90°$ we get the graph to the right.

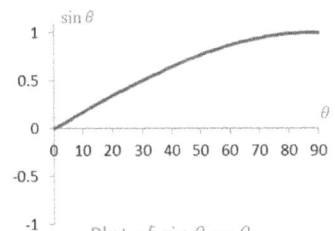

Plot of $\sin \theta$ vs θ.

Now let's put our sine function in the context of a coordinate system. If we look at the $x - y$ plane, we can see that there are four **quadrants**, or parts. We label each quadrant a capitalized roman numeral as shown to the right. If we draw a line from the origin and measure θ counterclockwise from the $+x$ −axis, then we can develop the following table:

Quadrant	x	y	θ
I	$x > 0$	$y > 0$	$0° < \theta < 90°$
II	$x < 0$	$y > 0$	$90° < \theta < 180°$
III	$x < 0$	$y < 0$	$180° < \theta < 270°$
IV	$x > 0$	$y < 0$	$270° < \theta < 360°$

If we align our original triangle so that θ is measured off of the x-axis, then

$$\text{ADJ}= x, \text{OPP}= y, \text{HYP}= r = \sqrt{x^2 + y^2},$$

and $\sin \theta = \frac{y}{r}$.

Note that whether or not x or y is negative, r is always positive. So, what happens to $\sin \theta$ when our triangle is in a quadrant other than I? In Quadrant II, $y > 0, r > 0$, so $\sin \theta > 0$. However, in Quadrants III and IV, $y < 0, r > 0$, so $\sin \theta < 0$. If we now plot $\sin \theta$ through a full 360°, we get the following plot:

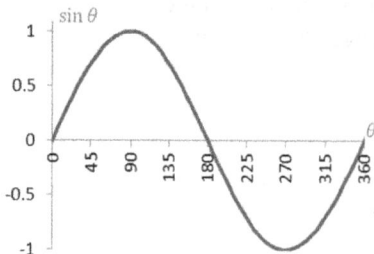

Plot of $\sin \theta$ vs θ.

For $\theta > 360°$, we go around in a circle again and repeat the pattern.

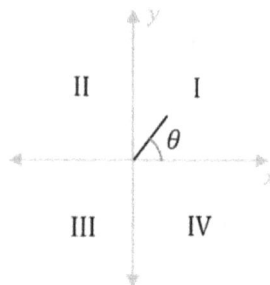

The $x - y$ plane.

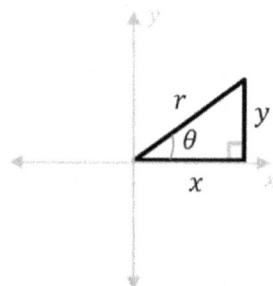

Right triangle in the $x - y$ plane.

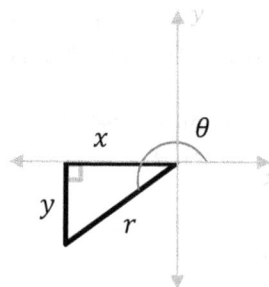

Right triangle in Quadrant III.

Taking the ratio of OPP/HYP is not the only way to predict the angle θ. We can take the ratio of any two sides of a right triangle. These ratios are called trigonometric ratios or **trigonometric functions**. Each ratio is given a special name: **sine**, **cosine**, and **tangent**.

Term	Notation	Equation
Sine	$\sin(\theta)$	$\sin(\theta) = \dfrac{OPP}{HYP} = \dfrac{y}{r}$
Cosine	$\cos(\theta)$	$\cos(\theta) = \dfrac{ADJ}{HYP} = \dfrac{x}{r}$
Tangent	$\tan(\theta)$	$\tan(\theta) = \dfrac{\sin(\theta)}{\cos(\theta)} = \dfrac{OPP}{ADJ} = \dfrac{y}{x}$

Below are plots of the ratios against the angle θ. Notice that $\tan\theta$ blows up (is undefined) when $x = 0$, i.e. when $\theta = 90°$ or $270°$.

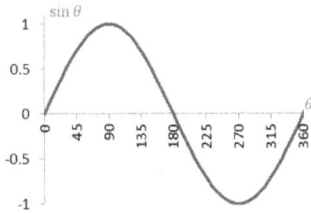

Plot of $\sin\theta$ vs θ.
$\sin\theta < 0$ in QIII, QIV.

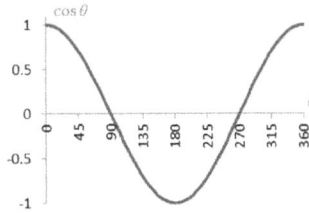

Plot of $\cos\theta$ vs θ.
$\cos\theta < 0$ in QII, QIII.

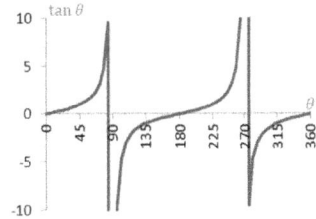

Plot of $\tan\theta$ vs θ.
$\tan\theta < 0$ in QII, QIV.

When you see a figure that is independent of a coordinate system, you can assume all sides have positive length no matter their orientation. For the commonly used angles, the following table is helpful and easy to remember:

	0°	30°	45°	60°	90°
$\sin\theta$	$\dfrac{\sqrt{0}}{2}$	$\dfrac{\sqrt{1}}{2}$	$\dfrac{\sqrt{2}}{2}$	$\dfrac{\sqrt{3}}{2}$	$\dfrac{\sqrt{4}}{2}$
$\cos\theta$	$\dfrac{\sqrt{4}}{2}$	$\dfrac{\sqrt{3}}{2}$	$\dfrac{\sqrt{2}}{2}$	$\dfrac{\sqrt{1}}{2}$	$\dfrac{\sqrt{0}}{2}$
$\tan\theta$	0	$(\sqrt{3})^{-1}$	$(\sqrt{3})^0$	$(\sqrt{3})^1$	∞

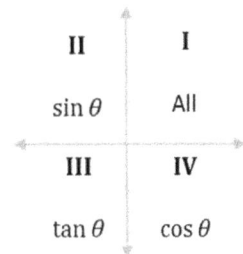

II	I
$\sin\theta$	All
III	IV
$\tan\theta$	$\cos\theta$

Functions >0 by quadrant. The phrase "ALL Students Take Calculus" helps recall which quadrants have which positive functions.

One final note before the examples: When you use your calculator to determine a trig value, make sure you know if your calculator is set to degrees or radians. As a reminder, radians are another measure of angle using π, the ratio of a circle's circumference to its diameter. There are 2π radians (or rad) in a circle just as there are 360°. To convert between radians and degrees, use the ratio $\pi/180°$.

$$45° = 45° \times \frac{\pi}{180°} = \frac{\pi}{4}$$

$$\frac{\pi}{3} = \frac{\pi}{3} \times \frac{180°}{\pi} = 60°$$

Degree-Radian Conversion.

Example

Find the value of the side indicated in the figure.

The two sides of interest are:

$$ADJ = 2"; \; HYP = r$$

The ratio of ADJ/HYP is the cosine of the angle:

$$\cos(30°) = \frac{ADJ}{HYP} = \frac{2}{r}$$

When $\theta = 30°$, $\cos(\theta) = \frac{\sqrt{3}}{2}$.

$$\frac{2}{r} = \frac{\sqrt{3}}{2} \rightarrow r = \frac{4}{\sqrt{3}} = 2.3".$$

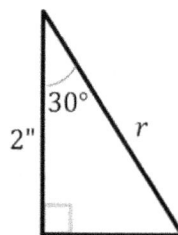

Right triangle for example.

Example

Use sine, cosine, and Pythagorean's theorem to determine the hypotenuse.

The sine of the angle is:

$$\sin 53° = \frac{4}{r} \rightarrow r = \frac{4}{\sin 53°}$$

Using a scientific calculator, when $\theta = 53°$, $\sin(\theta) = 0.8$, so $r = \frac{4}{0.8} = 5".$

The cosine of the angle is:

$$\cos 53° = \frac{3}{r} \rightarrow r = \frac{3}{\cos 53°}$$

Using a scientific calculator, when $\theta = 53°$, $\cos(\theta) = 0.6$, so $r = \frac{3}{0.6} = 5".$

Finally, using Pythagorean's theorem:

$$r = \sqrt{3^2 + 4^2} = \sqrt{25} = 5".$$

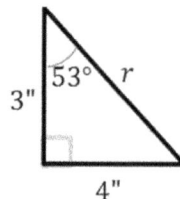

Right triangle for example.

Cosecant, Secant, & Cotangent:

There are other special ratios for a right triangle which can
be used in a same manner, though they're not as popular.
Also, once you know sine and cosine, you can derive the
rest:

Term	Notation	Equation
Cosecant	$\csc(\theta)$	$\csc(\theta) = \dfrac{1}{\sin(\theta)} = \dfrac{HYP}{OPP} = \dfrac{r}{y}$
Secant	$\sec(\theta)$	$\sec(\theta) = \dfrac{1}{\cos(\theta)} = \dfrac{HYP}{ADJ} = \dfrac{r}{x}$
Cotangent	$\cot(\theta)$	$\cot(\theta) = \dfrac{1}{\tan(\theta)} = \dfrac{ADJ}{OPP} = \dfrac{x}{y}$

The plots for these ratios are given below.

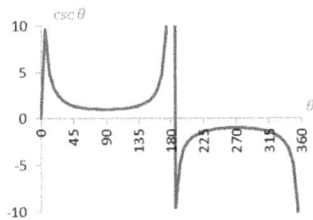

Plot of $\csc\theta$ vs θ. Plot of $\sec\theta$ vs θ. Plot of $\cot\theta$ vs θ.

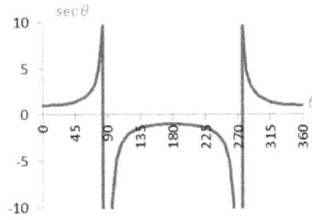

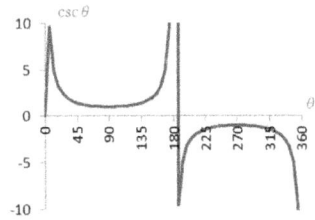

Example

Evaluate $\cot 60°$.

$$\cot 60° = \frac{1}{\tan 60°} = \frac{1}{\sqrt{3}} = \frac{\sqrt{3}}{3}.$$

Example

Simplify the expression: $\dfrac{\sec\theta}{\csc\theta}$.

$$\frac{\sec\theta}{\csc\theta} = \frac{1}{\cos\theta}\left(\frac{1}{\frac{1}{\sin\theta}}\right) = \frac{\sin\theta}{\cos\theta} = \tan\theta.$$

Chapter 11: Inverse Trigonometric Functions

Arc-Functions:

You can also work backward to figure out the value of the angle from its trigonometric value. In other words, you can determine an unknown angle of a right triangle simply by knowing the measure of any two sides. This reverse operation means taking an **inverse trigonometric function**. Specifically, you are taking the **inverse sine**, **inverse cosine**, or **inverse tangent**. It is also called taking the **arcsine**, **arccosine**, or **arctangent**.

Term	Notation	Equation
Arcsine	$asin(\theta)$ or $\sin^{-1}(\theta)$	$\theta = \sin^{-1}\left[\dfrac{OPP}{HYP}\right] = \sin^{-1}\left[\dfrac{y}{r}\right]$
Arccosine	$acos(\theta)$ or $\cos^{-1}(\theta)$	$\theta = \cos^{-1}\left[\dfrac{ADJ}{HYP}\right] = \cos^{-1}\left[\dfrac{x}{r}\right]$
Arctangent	$atan(\theta)$ or $\tan^{-1}(\theta)$	$\theta = \tan^{-1}\left[\dfrac{OPP}{ADJ}\right] = \tan^{-1}\left[\dfrac{y}{x}\right]$

Taking inverses of csc, sec, and cot is the same as for the other trig functions. Some students initially get confused by the "-1" notation for inverse. Remember that when placed over a function, like $f^{-1}(x)$, it means "undo the function." When placed over an expression that gives a numerical result, it means "divide into 1."

$$\sin^{-1}(\theta) \neq \frac{1}{\sin(\theta)}$$

$$\sin^{-1}(\theta) = asin(\theta) \qquad \text{gives an angle}$$

$$[\sin(\theta)]^{-1} = \frac{1}{\sin(\theta)} = \csc(\theta) \quad \text{gives a ratio}$$

This is also especially confusing because any other exponent is treated the opposite way:

$$\sin^2(\theta) = (\sin\theta)^2 \qquad \text{gives a ratio}$$

Because these are inverse functions, the following can be said:

$$\sin^{-1}(\sin\theta) = \theta; \quad \cos^{-1}(\cos\theta) = \theta;$$

and so on. In other words, the inverse of a function applied to itself gives the argument.

Example

Find the value of the angle indicated in the figure.

We know two sides: $OPP = 2"$; $HYP = 3"$.

The ratio of OPP/HYP is the sine of the angle:

$$\sin(\theta) = \frac{OPP}{HYP} = \frac{2}{3} = 0.67$$

Taking the inverse, we say:

$$\theta = \sin^{-1}\left(\frac{2}{3}\right)$$

Using a scientific calculator, $\theta = 42°$.

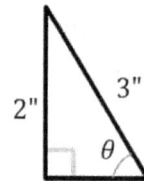
Right triangle for example.

Example

Find the value of the angle indicated in the figure.

We know two sides: $ADJ = 2"$; $OPP = 2"$.

The ratio of OPP/ADJ is the tangent of the angle:

$$\tan(\theta) = \frac{OPP}{ADJ} = \frac{2}{2} = 1$$

Taking the inverse, we say:

$$\theta = \tan^{-1}(1)$$

Using a scientific calculator, $\theta = 45°$.

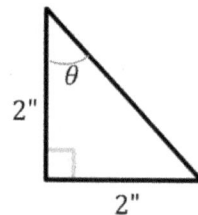
Right triangle for example.

Solving Trigonometric Equations:

Solving equations with trigonometric expressions is just like solving any equation you have learned thus far with one extra step. The final step is to take the inverse trig function and find all of the angles that satisfy the equation.

As you've seen, sine, cosine, and tangent are positive in two quadrants and negative in two quadrants. Therefore, an inverse trig function of a number should give two results (except for the arcsine or arccosine of ± 1 which are single-valued as per the plots on page 63) . To find these angles:

1. Ignore the sign of the argument to find the QI angle.
2. Use the sign to determine which quadrants hold the correct answers.
3. Measure the QI angle off of the x-axis.

Note: As before, in evaluating a drawn triangle independent of a coordinate system, you can assume all lengths are positive for the inverse trig function.

Example

Evaluate $\sin^{-1}\left(-\frac{1}{2}\right)$.

1. Our base angle is $\sin^{-1}\left(\frac{1}{2}\right) = 30°$.

2. The negative sign puts us in QIII and QIV.

3. Measuring off of the x-axis gives:

$$180 + 30 = 210°; \quad 360 - 30 = 330°.$$

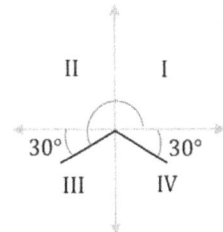

Two angle solutions.

In equation solving mode, the variable x is typically used instead of θ. This is just a notation and _not_ a reference to the x-axis or a particular side of a right triangle. Also, people work in radians rather than degrees. This is because radians are actual length ratios, so the units of an equation remain intact. For example:

$$4" \sin^{-1}\left(\frac{1}{2}\right) = 4" \cdot (0.52) = 2.1" \quad \text{NOT } 4" \cdot (30°).$$

To solve an equation with trig functions, make sure all of the functions are the same with the same argument. Then substitute a variable for the function to simplify the equation. Let's look at some examples.

Example

Solve for x: $4\sin x + 2 = 1$

Let's let $A = \sin x$:

$$4A + 2 = 1$$

$$A = -\frac{1}{4}$$

Since $A = \sin x$, $\sin x = -\frac{1}{4}$, and $x = \text{asin}\left(-\frac{1}{4}\right)$.

We know that $\sin \theta$ is negative in QIII and QIV, and $\text{asin}\left(+\frac{1}{4}\right) = 0.253 = 14.5°$ as measured from the x-axis. This means that there are two answers:

$$x = \pi + 0.253 = 3.4 \text{ rad (or } 194.5°)$$

and $x = 2\pi - 0.253 = 6.0$ rad (or $-14.5°$).

II   I

III   IV

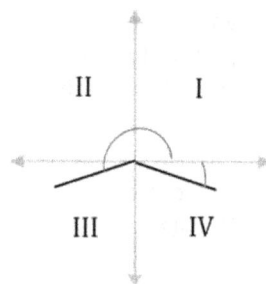

Two angle solutions.

Example

Solve for x: $\cos^2 x + \cos x = 2$

Let's let $A = \cos x$:

$$A^2 + A - 2 = 0$$

$$(A + 2)(A - 1) = 0$$

$$A = -2, 1$$

Since $A = \cos x$, $\cos x = \{-2, 1\}$.

We know from the graph of cosine that
$$-1 \leq \cos \theta \leq 1.$$

So, $\cos x = -2$ does not exist. As for $\cos x = 1$, that occurs when

$$x = \frac{\pi}{2} = 90° \text{ and } x = -\frac{\pi}{2} = -90°.$$

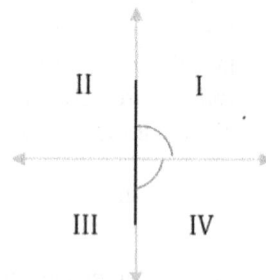

II   I

III   IV

Two angle solutions.

Chapter 12: Trigonometric Identities

Pythagorean Identities:

An **identity** is a relation that holds for any variable applied to it. You've already learned the **inverse identities**:

$\csc\theta = \dfrac{1}{\sin\theta}$	$\sec\theta = \dfrac{1}{\cos\theta}$	$\cot\theta = \dfrac{1}{\tan\theta}$
$\sin\theta = \dfrac{1}{\csc\theta}$	$\cos\theta = \dfrac{1}{\sec\theta}$	$\tan\theta = \dfrac{1}{\cot\theta}$

Based on the graphs of sine, cosine, and tangent (page 63), you can deduce the **negative angle identities**:

$\sin-\theta = -\sin\theta$	$\cos-\theta = \cos\theta$	$\tan-\theta = -\tan\theta$
$\csc-\theta = -\csc\theta$	$\sec-\theta = \sec\theta$	$\cot-\theta = -\cot\theta$

We'll develop a few more trig identities with the help of the **unit circle**. The unit circle is a circle with radius 1 centered at the origin.   If we pick points on the circle (x, y), then the following Pythagorean relation should always hold:

$$x^2 + y^2 = r^2 = 1.$$

And, the angle of the radius to the x-axis is defined with:

$$\cos\theta = \frac{x}{r} = x; \quad \sin\theta = \frac{y}{r} = y; \quad \tan\theta = \frac{y}{x}.$$

If we plug these relations into our Pythagorean relation we get:

$$y^2 + x^2 = 1$$

$$\sin^2\theta + \cos^2\theta = 1$$

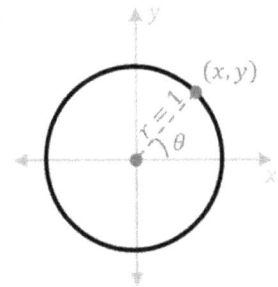

The unit circle.

Now let's divide both sides by $\cos^2 \theta$:

$$\frac{\sin^2 \theta}{\cos^2 \theta} + \frac{\cos^2 \theta}{\cos^2 \theta} = \frac{1}{\cos^2 \theta}$$

$$\left(\frac{\sin \theta}{\cos \theta}\right)^2 + 1 = \left(\frac{1}{\cos \theta}\right)^2$$

$$1 + \tan^2 \theta = \sec^2 \theta$$

If, instead, we divide both sides by $\sin^2 \theta$:

$$\frac{\sin^2 \theta}{\sin^2 \theta} + \frac{\cos^2 \theta}{\sin^2 \theta} = \frac{1}{\sin^2 \theta}$$

$$1 + \left(\frac{\cos \theta}{\sin \theta}\right)^2 = \left(\frac{1}{\sin \theta}\right)^2$$

$$1 + \cot^2 \theta = \csc^2 \theta$$

Because these equations were developed with Pythagorean relations on the unit circle, they are called the **Pythagorean identities**. They are summarized below.

$$\mathbf{\sin^2 \theta + \cos^2 \theta = 1}$$

$$\mathbf{1 + \tan^2 \theta = \sec^2 \theta}$$

$$\mathbf{1 + \cot^2 \theta = \csc^2 \theta}$$

Pythagorean identities.

Example

Verify that the Pythagorean identities hold for $-\theta$.

$$\sin^2(-\theta) + \cos^2(-\theta) = (-\sin \theta)^2 + (\cos \theta)^2$$
$$= \sin^2 \theta + \cos^2 \theta = 1$$

$$1 + \tan^2 -\theta = 1 + (-\tan \theta)^2$$
$$= 1 + \tan^2 \theta = \sec^2 \theta = (\sec - \theta)^2 = \sec^2 -\theta$$

$$1 + \cot^2 -\theta = 1 + (-\cot \theta)^2 = 1 + \cot^2 \theta$$
$$= \csc^2 \theta = (-\csc \theta)^2 = \csc^2 -\theta$$

Angle Sum & Difference Identities:

A common mistake among new students to trigonometry is to assume that $\sin(A + B) = \sin A + \sin B$. THIS IS NOT CORRECT! There are identities that give the sine, cosine, and tangent of the sum of two angles. We will state these formulas here, but a derivation of the first identity is given in Chapter 14. The **angle sum identities** are:

$$\sin(A + B) = \sin A \cos B + \sin B \cos A$$
$$\cos(A + B) = \cos A \cos B - \sin A \sin B$$
$$\tan(A + B) = \frac{\tan A + \tan B}{1 - \tan A \tan B}$$

Angle sum identities.

Example

Evaluate $\sin\left(\pi + \frac{\pi}{3}\right)$.

$$\sin\left(\pi + \frac{\pi}{3}\right) = \sin \pi \cos\frac{\pi}{3} + \sin\frac{\pi}{3}\cos \pi$$

$$= 0 \times \cos\frac{\pi}{3} + (-1) \times \sin\frac{\pi}{3} = -\sin 60° = -\frac{\sqrt{3}}{2}.$$

We can use the angle sum identities with the negative angle identities to develop the **angle difference identities**. For example:

$$\sin(A - B) = \sin A \cos(-B) + \sin(-B) \cos A$$

$$= \sin A \cos B - \sin B \cos A$$

Stating the all of the angle difference identities:

$$\sin(A - B) = \sin A \cos B - \sin B \cos A$$
$$\cos(A - B) = \cos A \cos B + \sin A \sin B$$
$$\tan(A - B) = \frac{\tan A - \tan B}{1 + \tan A \tan B}$$

Angle difference identities.

Double & Half Angle Identities:

We can use the angle sum identities again to develop the **double angle identities**.

> *Example*
>
> Find an expression for $\sin(2\theta)$.
>
> $$\sin(2\theta) = \sin(\theta + \theta) = \sin\theta\cos\theta + \sin\theta\cos\theta$$
>
> $$= 2\sin\theta\cos\theta$$

Stating all of the double angle identities:

$$\sin(2\theta) = 2\sin\theta\cos\theta$$

$$\cos(2\theta) = \cos^2\theta - \sin^2\theta$$

$$= 2\cos^2\theta - 1 = 1 - 2\sin^2\theta$$

$$\tan(2\theta) = \frac{2\tan\theta}{1-\tan^2\theta}$$

Double angle identities.

Finally, we can use the above to develop the half angle identities:

> *Example*
>
> Find an expression for $\cos(\theta/2)$.
>
> Treat θ as the double angle $\left(\frac{\theta}{2} + \frac{\theta}{2}\right)$:
>
> $$\cos(\theta) = \cos\left(2\left(\frac{\theta}{2}\right)\right) = 2\cos^2\left(\frac{\theta}{2}\right) - 1$$
>
> $$\cos\frac{\theta}{2} = \pm\sqrt{\frac{1+\cos\theta}{2}}\,.$$

The **half angle identities** are as follows:

$$\sin\frac{\theta}{2} = \pm\sqrt{\frac{1-\cos\theta}{2}}$$

$$\cos\frac{\theta}{2} = \pm\sqrt{\frac{1+\cos\theta}{2}}$$

$$\tan\frac{\theta}{2} = \pm\sqrt{\frac{1-\cos\theta}{1+\cos\theta}}$$

Half angle identities.

Product & Factoring Identities:

As you will discover in the problem sets, the remaining identities can also be derived. They will be stated here for convenience. As you can see, there are numerous trig identities. It is much easier to know the Pythagorean and angle sum identities and be able to derive the rest than to try to memorize all of them. The **product identities** are:

$$\sin A \cos B = \frac{1}{2}[\sin(A+B) + \sin(A-B)]$$

$$\cos A \cos B = \frac{1}{2}[\cos(A+B) + \cos(A-B)]$$

$$\sin A \sin B = \frac{1}{2}[\cos(A-B) - \cos(A+B)]$$

Product identities.

And, the **factoring identities** are:

$$\sin A + \sin B = 2\sin\left(\frac{A+B}{2}\right)\cos\left(\frac{A-B}{2}\right)$$

$$\sin A - \sin B = 2\cos\left(\frac{A+B}{2}\right)\sin\left(\frac{A-B}{2}\right)$$

$$\cos A + \cos B = 2\cos\left(\frac{A+B}{2}\right)\cos\left(\frac{A-B}{2}\right)$$

$$\cos A - \cos B = -2\sin\left(\frac{A+B}{2}\right)\sin\left(\frac{A-B}{2}\right)$$

Factoring identities.

Example

Derive $\sin A \cos B = \frac{1}{2}[\sin(A + B) + \sin(A - B)]$

Looking at the right side of the equation, it would suggest we add the equations for sine sum and sine difference:

$$\sin(A + B) = \sin A \cos B + \sin B \cos A$$
$$+\sin(A - B) = \sin A \cos B - \sin B \cos A$$
$$\sin(A + B) + \sin(A - B) = 2 \sin A \cos B$$

$$\sin A \cos B = \frac{1}{2}[\sin(A + B) + \sin(A - B)]$$

Example

Derive $\sin A + \sin B = 2 \sin\left(\frac{A+B}{2}\right) \cos\left(\frac{A-B}{2}\right)$

Let's work on expanding the right side using the product identity:

$$\sin A' \cos B' = \frac{1}{2}[\sin(A' + B') + \sin(A' - B')]$$

with the following assignments:

$$A' \equiv \frac{A+B}{2}; \quad B' \equiv \frac{A-B}{2}$$

$$2 \sin\left(\frac{A+B}{2}\right) \cos\left(\frac{A-B}{2}\right)$$

$$= 2\left(\frac{1}{2}\right)\left[\sin\left(\frac{A+B}{2} + \frac{A-B}{2}\right) + \sin\left(\frac{A+B}{2} - \frac{A-B}{2}\right)\right]$$

$$= \sin\frac{2A}{2} + \sin\frac{2B}{2} = \sin A + \sin B$$

Chapter 13: Laws of Sines & Cosines

Law of Sines:

The laws in this chapter help us determine sides and angles for triangles that do not have right angles.

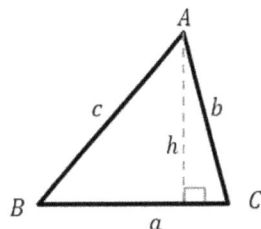

Triangle ABC with base a.

The area of $\triangle ABC$ shown to the right is:

$$Area = \frac{1}{2}\,base \times height.$$

Here $base = a$, and using trigonometry, we can say that

$$height = h = b \sin C.$$

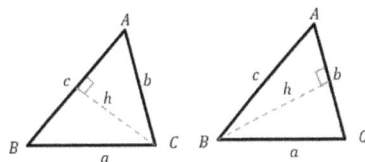

Triangle ABC with different bases.

Therefore, $Area = \frac{1}{2}ab \sin C$.

We can also find the cases where c or b is the base. No matter which way we do it, the area should always be the same, so we get the equality:

$$\frac{1}{2}ab \sin C = \frac{1}{2}ac \sin B = \frac{1}{2}bc \sin A$$

Multiplying through by $2/(abc)$ we get the **Law of Sines**:

$$\frac{\sin A}{a} = \frac{\sin B}{b} = \frac{\sin C}{c}$$

In other words, the ratio of the sine of any angle to the length of the opposite side is the same within any triangle.

It is best to use the Law of Sines when you know at least two angles *and* at least one side. Specifically:
- AAS – Angle Angle Side: Two angles and the non-included side
- ASA – Angle Side Angle: Two angles and the side between them

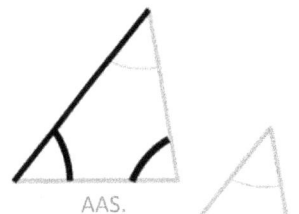

AAS.

ASA.

You can use the Law of Sines for the SSA (Side Side Angle) case, but you may get up to two viable answers.

SSA.

Example

Find the lengths of the other two sides:

We have an AAS case. We set up our Law of Sines as follows:

$$\frac{\sin 60°}{a} = \frac{\sin(180-45-60)°}{b} = \frac{\sin 45°}{4}$$

$$a = \frac{4 \sin 60°}{\sin 45°} = 4\left(\frac{\sqrt{3}}{2}\right)\left(\frac{2}{\sqrt{2}}\right) = 4.9"$$

$$b = \frac{4 \sin 75°}{\sin 45°} = 5.5"$$

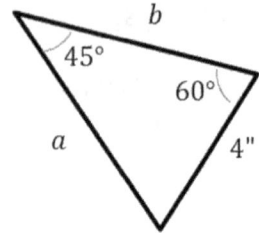

Triangle for example.

Example

Find the lengths of the other two sides:

We have an ASA case. We set up our Law of Sines as follows:

$$\frac{\sin 70°}{a} = \frac{\sin 60°}{b} = \frac{\sin(180-70-60)°}{3}$$

$$a = \frac{3 \sin 70°}{\sin 50°} = 3.7"$$

$$b = \frac{3 \sin 60°}{\sin 50°} = 3.4"$$

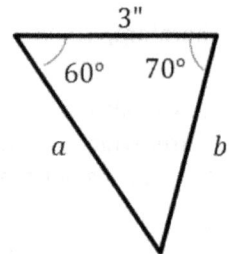

Triangle for example.

Example

Find the length of the third side:

We have an SSA case, so we expect up to two answers. We set up our Law of Sines as follows:

$$\frac{\sin A}{a} = \frac{\sin B}{7} = \frac{\sin 50°}{6}$$

$$B = \sin^{-1}\left[\frac{7}{6}\sin 50°\right] = \sin^{-1} 0.766 = 50°, 130°$$

Sine is positive in QI and QII. Looking at the figure, we have an acute triangle, so we assume $B = 50°$.

$$a = \frac{6 \sin(180-50-50)°}{\sin 50°} = 7.7"$$

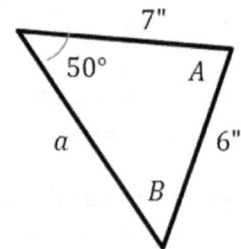

Triangle for example.

Law of Cosines:

Now let's use the Pythagorean theorem on parts of $\triangle ABC$. To get started, we drop a perpendicular to side a and split it into:
$$a = c \cos B + b \cos C.$$

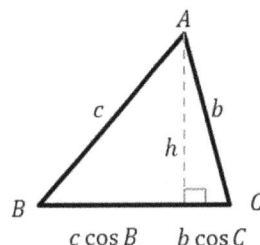

Triangle ABC with base a.

Multiplying through by a, we get:
$$a^2 = ac \cos B + ab \cos C.$$

Splitting up side b and multiplying through by b we get:
$$b^2 = bc \cos A + ab \cos C.$$
Doing a similar thing with side c we get:
$$c^2 = ac \cos B + bc \cos A.$$

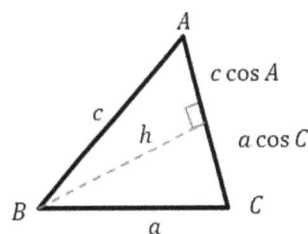

Triangle ABC with base b.

Now let's evaluate $a^2 + b^2 - c^2$:
$$a^2 + b^2 - c^2$$
$$= ac \cos B + ab \cos C + bc \cos A$$
$$+ ab \cos C - ac \cos B - bc \cos A$$
Simplifying, we get:
$$c^2 = a^2 + b^2 - 2ab \cos C.$$

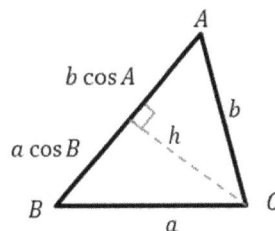

Triangle ABC with base c.

We can also evaluate $a^2 - b^2 + c^2$ and $-a^2 + b^2 + c^2$: The results are three versions of the **Law of Cosines**:

$$c^2 = a^2 + b^2 - 2ab \cos C$$
$$b^2 = a^2 + c^2 - 2ac \cos B$$
$$a^2 = b^2 + c^2 - 2bc \cos A$$

It is best to use the Law of Cosines when you know at least two sides. Specifically:
- SSA – Two sides and an adjacent angle
- SAS – Side Angle Side: Two sides and the angle between
- SSS– Side Side Side: All three sides

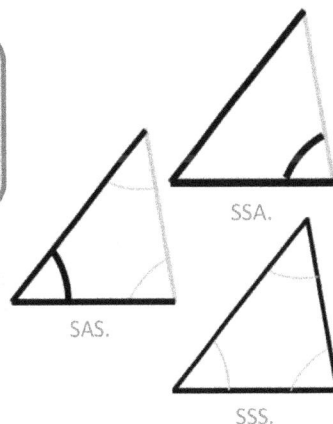

SSA.

SAS.

SSS.

79

> **Example**
>
> Find the length of the third side:
>
> We have an SSA case. We set up our Law of Cosines as follows:
>
> $$c^2 = a^2 + b^2 - 2ab \cos C$$
>
> $$5^2 = a^2 + 3^2 - 6a \cos 60°$$
>
> $$a^2 - 3a - 16 = 0$$
>
> $$a = \frac{3 \pm \sqrt{9 - 4(-16)}}{2}$$
>
> $$a = -2.77, 5.77$$
>
> Only the positive length makes physical sense, so $a = 5.77"$.

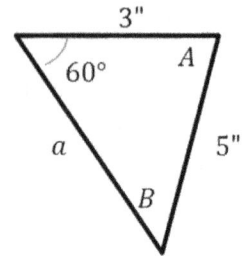

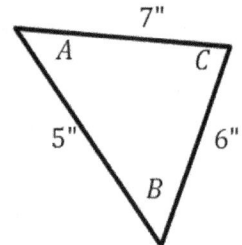

Triangle for example.

> **Example**
>
> Find the angles:
>
> We have a SSS case. We set up our Law of Cosines as follows (you can choose any form):
>
> $$c^2 = a^2 + b^2 - 2ab \cos C$$
>
> $$5^2 = 6^2 + 7^2 - 84 \cos C$$
>
> $$C = \cos^{-1} 0.714 = 44.4°, -44.4°$$
>
> We will concern ourselves with the positive angle since we don't have a coordinate system. We can now use the Law of Sines to find A:
>
> $$\frac{\sin A}{6} = \frac{\sin 44.4°}{5}$$
>
> $$A = \sin^{-1} \left[\frac{6}{5} \sin 44.4°\right] = 57°, 123°.$$
>
> By the figure, we assume the acute angle, so $A = 57°$. The sum of angles for a triangle gives us the third angle:
>
> $$B = 180 - A - C = 78.6°.$$

Triangle for example.

Chapter 14: Hyperbolic Trig Functions

Euler's Identity:

There is an alternate way to write sine and cosine using exponentials and imaginary numbers. These formulas stem from **Euler's identity**:

$$e^{i\theta} = \cos\theta + i\sin\theta$$

For negative angles we have a similar expression:

$$e^{-i\theta} = \cos-\theta + i\sin-\theta = \cos\theta - i\sin\theta$$

With some manipulation, we can arrive at the following:

$$\sin\theta = \frac{1}{2i}\left(e^{i\theta} - e^{-i\theta}\right); \cos\theta = \frac{1}{2}\left(e^{i\theta} + e^{-i\theta}\right).$$

Let's use the above expressions to derive the $\sin(A + B)$ identity:

$$\sin(A + B) = \frac{1}{2i}\left(e^{i(A+B)} - e^{-i(A+B)}\right)$$

$$= \frac{1}{2i}\left(e^{iA}e^{iB} - e^{-iA}e^{-iB}\right)$$

Plugging Euler's identity back in and doing lots of Algebra:

$$= \frac{1}{2i}[(\cos A + i\sin A)(\cos B + i\sin B)$$
$$- (\cos A - i\sin A)(\cos B - i\sin B)]$$

$$= \frac{1}{2i}[\cos A\cos B + i\sin B\cos A + i\sin A\cos B$$
$$- \sin A\sin B - \cos A\cos B + i\sin B\cos A$$
$$+ i\sin A\cos B + \sin A\sin B]$$

$$= \frac{1}{2i}[2i\,[\sin A\cos B + \sin B\cos A]$$

$$= \sin A\cos B + \sin B\cos A.$$

Hyperbolic Trig Functions:

The "common" trig functions we've learned about thus far are actually called the **circular trigonometric functions**. This is because of the relation:

$$\sin^2\theta + \cos^2\theta = 1.$$

For the unit circle, $\sin^2\theta = y^2$; $\cos^2\theta = x^2$. Plugging back into the Pythagorean identity we get:

$$x^2 + y^2 = 1$$

or, the equation of a circle centered at the origin with radius 1.

The two main **hyperbolic trigonometric functions** are the **hyperbolic sine**, sinh, and the **hyperbolic cosine**, cosh. They are pronounced sinch and cosh, for short. These functions are defined as follows:

$$\sinh\theta = \frac{e^\theta - e^{-\theta}}{2}; \quad \cosh\theta = \frac{e^\theta + e^{-\theta}}{2}$$

Let's look at what happens when we square both terms and take their difference:

$$\cosh^2\theta - \sinh^2\theta$$

$$= \frac{e^{2\theta} + 2e^\theta e^{-\theta} + e^{-2\theta}}{4} - \frac{e^{2\theta} - 2e^\theta e^{-\theta} + e^{-2\theta}}{4}$$

$$= \frac{4e^{\theta-\theta}}{4} = e^0 = 1$$

So, we have the relation:

$$\cosh^2\theta - \sinh^2\theta = 1$$

If you plot $\cosh\theta$ versus $\sinh\theta$ for $\cosh^2 x - \sinh^2 x = 1$, you get something that looks like $x^2 - y^2 = 1$ which is the equation for a hyperbola. This is the reasoning behind the term hyperbolic trig function.

The other hyperbolic functions can be derived using similar definitions as circular trig relations:

$$\tanh\theta = \frac{\sinh\theta}{\cosh\theta} = \frac{e^\theta - e^{-\theta}}{e^\theta + e^{-\theta}}$$

$$\text{csch}\,\theta = \frac{1}{\sinh\theta} = \frac{2}{e^\theta - e^{-\theta}}$$

$$\text{sech}\,\theta = \frac{1}{\cosh\theta} = \frac{2}{e^\theta + e^{-\theta}}$$

$$\coth\theta = \frac{1}{\tanh\theta} = \frac{e^\theta + e^{-\theta}}{e^\theta - e^{-\theta}}$$

Two additional relations that are commonly used are listed below. A more complete summary of hyperbolic trig identities is given in Appendix E.

$$1 - \tanh^2\theta = \text{sech}^2\theta$$

$$\coth^2\theta - 1 = \text{csch}^2\theta$$

Example

Verify that $1 - \tanh^2\theta = \text{sech}^2\theta$

$$1 - \tanh^2\theta = 1 - \frac{\sinh^2\theta}{\cosh^2\theta}$$

$$= \frac{\cosh^2\theta - \sinh^2\theta}{\cosh^2\theta}$$

$$= \frac{1}{\cosh^2\theta} = \text{sech}^2\theta$$

Inverse Hyperbolic Trig Functions:

As you might expect, the hyperbolic functions have inverses called **inverse hyperbolic functions**. The three principle ones are:

$$\sinh^{-1}\theta = \ln\left(\theta + \sqrt{\theta^2 + 1}\right)$$

$$\cosh^{-1}\theta = ln(\theta + \sqrt{\theta^2 - 1})$$

$$\tanh^{-1}\theta = \frac{1}{2}\ln\frac{1+\theta}{1-\theta}$$

Example

Verify that $\sinh^{-1}(\sinh\theta) = \theta$.

$$\sinh^{-1}(\sinh\theta) = \ln\left(\sinh\theta + \sqrt{(\sinh\theta)^2+1}\right)$$

$$= \ln(\sinh\theta + \cosh\theta)$$

$$= \ln\left(\frac{e^\theta - e^{-\theta} + e^\theta + e^{-\theta}}{2}\right)$$

$$= \ln\left(\frac{2e^\theta}{2}\right) = \ln(e^\theta) = \theta.$$

Example

Verify that $\sinh(\sinh^{-1}\theta) = \theta$.

$$\sinh(\sinh^{-1}\theta)$$

$$= \frac{1}{2}\left[e^{\ln(\theta+\sqrt{\theta^2+1})} - e^{-\ln(\theta+\sqrt{\theta^2+1})}\right]$$

$$= \frac{1}{2}\left[\theta + \sqrt{\theta^2 + 1} - \frac{1}{\theta+\sqrt{\theta^2+1}}\right] \qquad \text{Rationalize denominator}$$
$$\text{with } \frac{\theta-\sqrt{\theta^2+1}}{\theta-\sqrt{\theta^2+1}}.$$

$$= \frac{1}{2}\left[\theta + \sqrt{\theta^2 + 1} - \frac{\theta-\sqrt{\theta^2+1}}{\theta^2-(\theta^2+1)}\right]$$

$$= \frac{1}{2}\left[\theta + \sqrt{\theta^2 + 1} - \frac{\theta-\sqrt{\theta^2+1}}{-1}\right] = \frac{1}{2}(2\theta) = \theta.$$

Appendices

A: Course Summary
B: Problem Sets
C: Solutions to Problem Sets
D: Trigonometric Identities
E: Hyperbolic Trig Identities

Appendix A: Course Summary

1   **Review** of Algebra I:
       Numbers & Fractions
       Order of Operations & Operation Properties
       Exponents
       Binomial Expressions

2   **Numbers & Functions:**
       Imaginary ($i = \sqrt{-1}$) and Complex ($a + bi$) Numbers:
         Real and Imaginary Operators: $\mathfrak{Re}\{a + bi\} = a$; $\mathfrak{Im}\{a + bi\} = b$
         Complex Conjugate: $(a + bi)^* = a - bi$
         Solving the **Quadratic Equation** for Complex Roots

3   **Functions:**
       Vertical Line Test
       Composite Functions: $g \circ f(x) = g(f(x))$
       Inverse Functions of Functions Return the Argument: $f^{-1}(f(x)) = x$

4   **Logarithms & Exponentials**
       $\log_b y = x \rightarrow y = b^x$; $\ln y = x \rightarrow y = e^x$
       Properties & Conversions **p 23**
       Factorials

5   **Graphing:**
       Review & Definitions
       Common Functions:
         Linear, Quadratic, Cubic, Logarithmic, Exponential

6   **Lines:**

$$distance = \sqrt{(x_2 - x_1)^2 + (y_2 - y_1)^2}$$

$$midpoint = \left(\frac{x_1 + x_2}{2}, \frac{y_1 + y_2}{2}\right)$$

$$slope = m = \frac{\Delta y}{\Delta x} = \frac{y_2 - y_1}{x_2 - x_1}$$

Point-Slope Form: $y - b = m(x - a)$

Slope-Intercept Form: $y = mx + b$

7 **Circles & Conics**: Summary on **p 45**

   Circles: $x^2 + y^2 = r^2$

   Parabolas: $x^2 = 4py$; $y^2 = 4px$

   Ellipses: $\frac{x^2}{a^2} + \frac{y^2}{b^2} = 1$; $\frac{y^2}{a^2} + \frac{x^2}{b^2} = 1$

   Hyperbolas: $\frac{x^2}{a^2} - \frac{y^2}{b^2} = 1$; $\frac{y^2}{a^2} - \frac{x^2}{b^2} = 1$

8 Solving Systems of Independent Equations:
   Combination, Substitution, Graphing

9 **Congruent Transformations**:
   Translations:
   $$x' = x - h; \quad y' = y - k$$
   Reflections
   About x-axis: $y' = -y$
   About y-axis: $x' = -x$
   Rotations:
   $$x' = x\cos\theta + y\sin\theta; \quad y' = -x\sin\theta + y\cos\theta.$$

Trigonometry:
10   **Circular Trig** Functions
11   **Inverse** Functions & **Quadrants**
12   **Identifies**: Summary in **Appendix D**
         Inverse, Negative Angle, Pythagorean, Angle Sum & Difference,
         Double Angle, Half Angle, Product, Factoring
13   Law of Sines: ASA, AAS, SSA

         $$\frac{\sin A}{a} = \frac{\sin B}{b} = \frac{\sin C}{c}$$
         Law of Cosines: SAS, SSS, SSA

         $$c^2 = a^2 + b^2 - 2ab\cos C$$
14   **Hyperbolic Trig** Functions
      Inverse Functions
      Expressions with **Exponentials**
      Identities: Summary in **Appendix E**

Chapter correlations are given in large grey font.

Appendix B: Problem Sets

Chapter 1

Evaluate:

1.1: $5 + (-9) - (-4)$

1.2: $2 + 8 - (-3)$

1.3: -2×8

1.4: $-3 \times -4 \times -2$

1.5: $\frac{5}{12} - \frac{4}{20}$

1.6: $\frac{2}{3} - \frac{3}{4} + \frac{1}{6}$

1.7: $-\frac{3}{2} \times \frac{2}{5}$

1.8: $\frac{3}{4} \times \frac{1}{12} \times \frac{2}{3}$

1.9: $5 \times (2 - 3)$

1.10: $3 \times \left(4 - (8 + 2)\right)$

1.11: $(2^2 + 3)^2$

1.12: $\left(\frac{18}{(4-2)}\right)\left(\frac{(-2)^2}{6}\right)$

1.13: $x^2 = 2 - x$

1.14: $-x^2 = 2x - 3$

1.15: $x^2 = 5 - 4x$

1.16: $x^2 + \frac{7x}{2} = 2$

Chapter 2

Simplify:

2.1: i^{12}

2.2: $(-i)^4$

2.3: $\frac{2i}{i^3}$

2.4: $-4i^3 \times i^2$

2.5: $(i+1) - (2i-1)$

2.6: $\frac{i+1}{i-1}$

2.7: $\frac{(1-i)(1+3i)}{3-2i}$

2.8: $2i - \frac{3+3i}{2+2i}$

Solve for x:

2.9: $x^2 + 4x + 6$

2.10: $x^2 + 2x + 4$

2.11: $x^2 - x + 9$

2.12: $2x^2 - 3x + 2$

Find the polynomial with the given roots:

2.13: $(2+i),\ (2-i)$

2.14: $(3+2i),\ (3-2i)$

2.15: $(1+4i),\ (1-4i)$

2.16: $(-2+4i),\ (-2-4i)$

Chapter 3

Determine if the following can be expressed as functions:

3.1: $4x - 2y = 6$

3.2: $y = x^2$

3.3: $x = y^2$

3.4: $y = \begin{cases} x & x \in [0, \infty) \\ -1 & x \in (-\infty, 0] \end{cases}$

Evaluate $(f \circ g)(x)$ and $(g \circ f)(x)$:

3.5: $f(x) = 2x$;
$g(x) = 5x + 8$

3.6: $f(x) = x + 1$;
$g(x) = 3x - 4$

3.7: $f(x) = x + 3$;
$g(x) = 4x - 6$

3.8: $f(x) = \sqrt{x}$;
$g(x) = x^2 + 4x + 4$

3.9: $f(x) = 3x^2$;
$g(x) = 2x + 4$

3.10: $f(x) = x^2 - 2x$;
$g(x) = x - 3$

3.11: $f(x) = x^2 - 3x + 9$;
$g(x) = x + 2$

3.12: $f(x) = (3 + 2i)x$;
$g(x) = \Re\{4x\}$

Find the inverse function:

3.13: $f(x) = 3x + 4$

3.14: $f(x) = x^3 + 3$

3.15: $f(x) = 3\sqrt{x} - 5$

3.16: $f(x) = (1 - i)x + 2$

Chapter 4

Write the following in exponential form:

4.1: $\log_5 2 = x$ **4.2:** $\log_3 x = 8$

4.3: $\log x = 20$ **4.4:** $\ln 4 = x$

Solve for x:

4.5: $25 = \dfrac{1}{5^x}$ **4.6:** $16^x = 4$

4.7: $3^{(x^2+2x)} = 27$ **4.8:** $32^x(2^{3x}) = 16$

4.9: $5^{2x} = 6$ **4.10:** $4^{(x+1)} = 3^x$

4.11: $5^{(x-3)} = \dfrac{1}{2^x}$ **4.12:** $3^{2x} = \dfrac{1}{16^{(x+1)}}$

Evaluate:

4.13: $\log_2 40$ **4.14:** $\log_5 100$

4.15: $\log_7 8$ **4.16:** $\log_{1/4} 3$

Chapter 5

Plot the following:

5.1: $4y + 8x = 16$ **5.2:** $2y - 6 = 12x$

5.3: $5y = 15\sqrt{x} + 40$ **5.4:** $3y = x^2 - 9$

5.5: $y = 3x^3 - 2x$ **5.6:** $y = x^2 - 5x^3$

5.7: $y = x - 4x^3$ **5.8:** $y = x^3 - 3x^2 - 4x - 5$

5.9: $y = \log(10x + 2)$ **5.10:** $y = \ln(x^2 + 2)$

5.11: $y = x^2 \ln x$ **5.12:** $y = \log_2 x$

5.13: $y = 3^x + x^3$ **5.14:** $y = 4^{(x+1)}$

5.15: $y = x^x$ **5.16:** $y = 2^{-x} + 2^{x^2}$

Chapter 6

Find the distance, midpoint, and slope for the following pairs of points:

6.1: $(1,2); (3,4)$ **6.2:** $(-5,2); (3,8)$

6.3: $(-1,10); (-2,6)$ **6.4:** $(1,0); (-1,2)$

6.5: $(-5,-5); (2,2)$ **6.6:** $(4,-3); (8-1)$

Find the y-intercept for the line:

6.7: For Problem 6.1:
     $(1,2); (3,4)$

6.8: For Problem 6.2:
     $(-5,2); (3,8)$

6.9: For Problem 6.3:
     $(-1,10); (-2,6)$

6.10: For Problem 6.4:
     $(1,0); (-1,2)$

6.11: $2y + 4x = 6$ **6.12:** $2(3 - y) = 6x$

6.13: $4 - y = 5x + 2$ **6.14:** $x + 5 = 9 - y$

Given the line $y = 3x - 1$ find the following:

6.15: A parallel line through the point $(2,1)$

6.16: A perpendicular line through the point $(2,1)$

Chapter 7

Develop the equation for the figure indicated:

7.1: A circle of radius 4 centered on the origin.

7.2: A circle of diameter 4 centered on the point $(-1,5)$.

7.3: A parabola that opens to the right withwith $|p| = 3$.

7.4: A parabola that opens downward withwith $|p| = 2$.

7.5: An ellipse with major and minor axes endpoints at $(\pm 4,0), (0,\pm 2)$.

7.6: An ellipse with a major axis from $(0,\pm 5)$ and foci at $(0,\pm 4)$.

7.7: A hyperbola with foci at $(\pm\sqrt{13}, 0)$ and vertices at $(\pm 3,0)$.

7.8: A hyperbola with vertices at $(0,\pm 12)$ and asymptotes of $y = \pm 6x$.

Sketch the graph of the equation given:

7.9: $(x-1)^2 + (y+1)^2 = 4$

7.10: $(x+2)^2 + (y-3)^2 = 1$

7.11: $x^2 = 16y$

7.12: $y^2 = -8x$

7.13: $\dfrac{x^2}{16} + \dfrac{y^2}{4} = 1$

7.14: $\dfrac{y^2}{25} + \dfrac{x^2}{9} = 1$

7.15: $\dfrac{x^2}{9} - \dfrac{y^2}{4} = 1$

7.16: $\dfrac{y^2}{16} - \dfrac{x^2}{4} = 1$

Chapter 8

Solve by Combination:

8.1: $2y = 2x + 3; \ y = 4x$ **8.2:** $y = x - 2; \ 2y = 3x + 2$

8.3: $3y = x^2 + 8; \ y = 4 - x$ **8.4:** $4y = x^2 + 7; \ y = x + 1$

8.5: $y = x + 3; \ 3y = 2x^2 + 10$ **8.6:** $y = -x; \ 2y = x^2 - 3$

Solve by Substitution:

8.7: $y = x + 1; \ 4y = 5x - 1$ **8.8:** $y = 3x + 3; \ 2y = x + 10$

8.9: $y = 4x - 2; \ 3y = 4x^2 + 2$ **8.10:** $y = 10x + 4; \ \frac{y}{2} = 4x + 3$

8.11: $y = 2x^2; \ 3y = 5x + 1$ **8.12:** $y = x - 1; \ 2y = x^2 - 1$

Solve by Graphing Solutions:

8.13: $y = 3x - 2; \ 2y = 4x + 5$ **8.14:** $y = x + 2; \ y = x^2$

8.15: $y = 4x - 1; \ y = x^3 - 3$ **8.16:** $y = 3x + 5; \ y = x^2 + 1$

Chapter 9

Determine the equations for the following translated conics:

9.1:

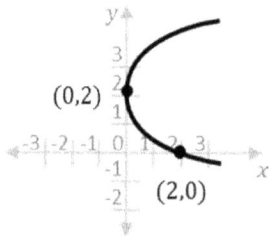

9.2:

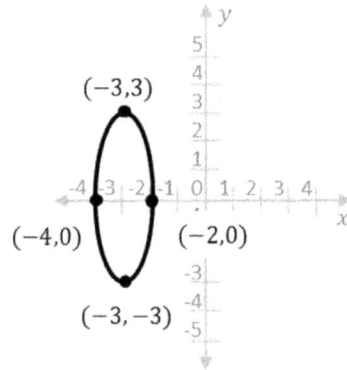

9.3:

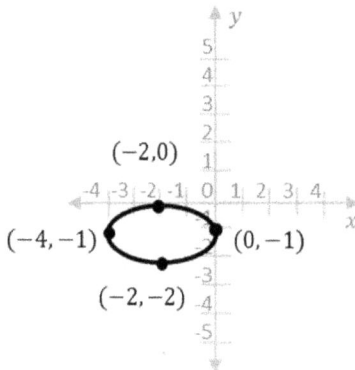

9.4:

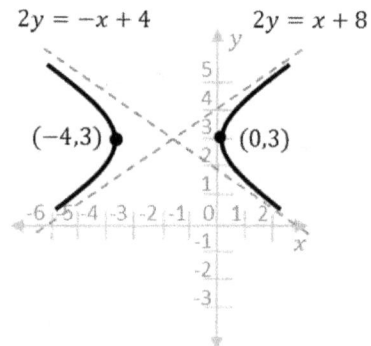

Write the equation and sketch the transformations under the figure as indicated:

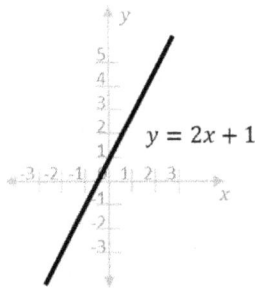

$y = 2x + 1$

$(x + 2)^2 + (y - 3)^2 = 1$

9.5: Reflect about the x-axis

9.6: Rotate $90°$

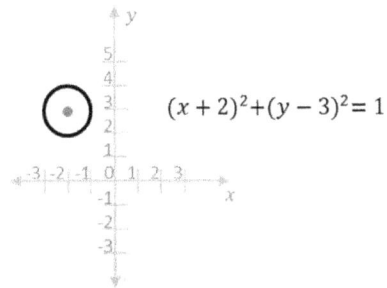

9.7: Translate 4 units down and 1 unit right

9.8: Reflect about the y-axis

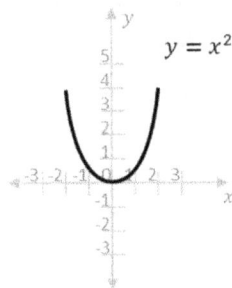

$y = x^2$

$$\frac{x^2}{4} + (y - 1)^2 = 1$$

9.9: Translate 2 units up and 1 unit left

9.10: Rotate $90°$

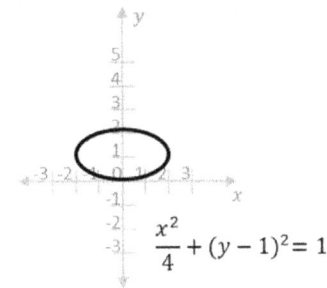

9.11: Reflect about the x-axis

9.12: Rotate $60°$

$$\frac{(y - 3)^2}{4} + (x - 1)^2 = 1$$

$$\frac{x^2}{9} - \frac{y^2}{4} = 1$$

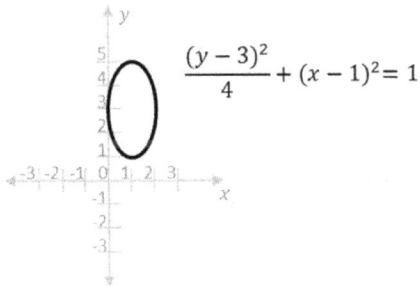

9.13: Translate 3 units down and 2 units left

9.14: Reflect about the y-axis

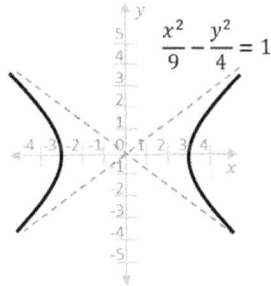

9.15: Translate 2 units up and 1 unit right

9.16: Rotate $45°$

Chapter 10

Find the side indicated:

10.1:

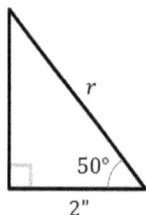

10.2:

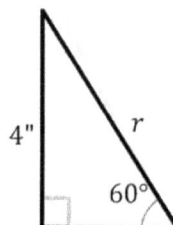

10.3:

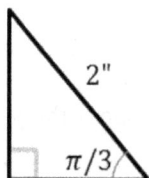

10.4:

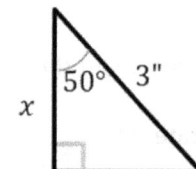

10.5:

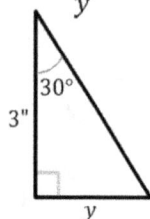

10.6:

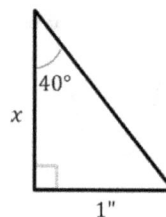

10.7:

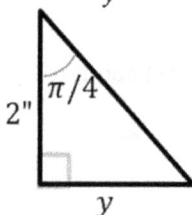

10.8:

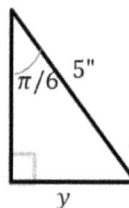

Evaluate:

10.9: $\csc 30°$

10.10: $\cot\left(\frac{\pi}{4}\right)$

10.11: $\sec\left(\frac{\pi}{6}\right)$

10.12: $\csc 45°$

Simplify:

10.13: $\dfrac{\csc\theta}{\cot\theta}$

10.14: $\tan\theta \, \csc\theta$

10.15: $\cot\theta \sin\theta$

10.16: $\cos\theta \, \csc\theta$

Chapter 11

Find the missing angles:

11.1:

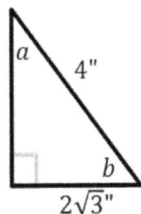

11.2:

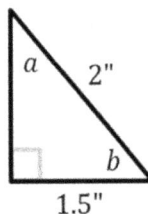

11.3:

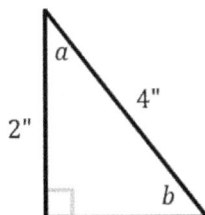

11.4:

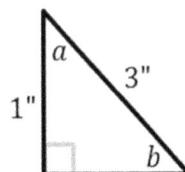

11.5:

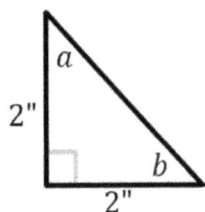

11.6:

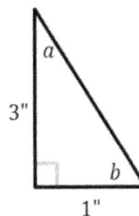

Evaluate:

11.7: $\sin^{-1} \frac{\sqrt{2}}{2}$

11.8: $\cos^{-1}\left(-\frac{\sqrt{3}}{2}\right)$

11.9: $\sin^{-1} 0$

11.10: $\tan^{-1}(-\sqrt{3})$

Solve for x:

11.11: $2 \sin x + 1 = 3$

11.12: $2 \cos 2x = 1$

11.13: $\sin^2 x = 5 - 4 \sin x$

11.14: $-\tan^2 x = 2 \tan x - 3$

11.15: $\cos^2 x + \frac{7}{2} \cos x = 2$

11.16: $\tan^2 x = 1 + 2 \tan x$

Chapter 12

Evaluate:

12.1: $\cos -45°$

12.2: $\tan -60°$

12.3: $\sin 135°$

12.4: $\cos 120°$

Prove the following identities:

12.5: $\cos(A - B)$
$= \cos A \cos B + \sin A \sin B$

12.6: $\cos 2\theta = \cos^2 \theta - \sin^2 \theta$

12.7: $\tan 2\theta = \frac{2 \tan \theta}{1 - \tan^2 \theta}$

12.8: $\sin \frac{\theta}{2} = \pm \sqrt{\frac{1 - \cos \theta}{2}}$

12.9: $\cos A \cos B$
$= \frac{1}{2}[\cos(A + B) + \cos(A - B)]$

12.10: $\sin A \sin B$
$= \frac{1}{2}[\cos(A - B) - \cos(A + B)]$

12.11: $\sin A - \sin B$
$= 2 \sin\left(\frac{A-B}{2}\right) \cos\left(\frac{A+B}{2}\right)$

12.12: $\cos A + \cos B$
$= 2 \cos\left(\frac{A+B}{2}\right) \cos\left(\frac{A-B}{2}\right)$

Solve for x:

12.13: $2 \sin x \cos x = 1$

12.14: $4x + 2 \sin^2 \frac{x}{2} + \cos x = 3$

12.15: $3 \cos x = 4\sin^2 x - 3$

12.16: $\sin x = \cos 2x$

Chapter 13

Find the missing quantities:

13.1:

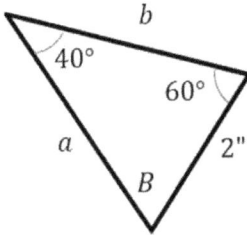

13.2:

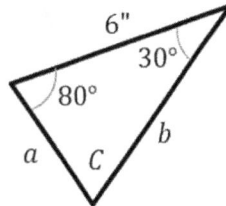

13.3:

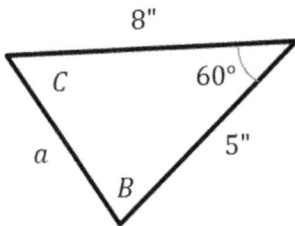

13.4:

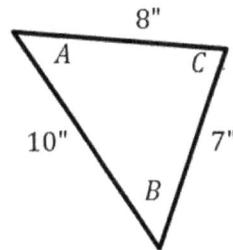

13.5:

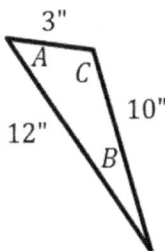

13.6:

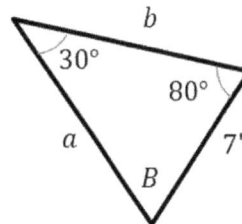

13.7:

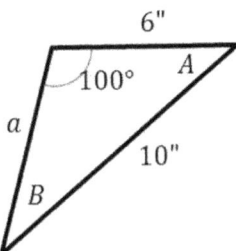

13.8:

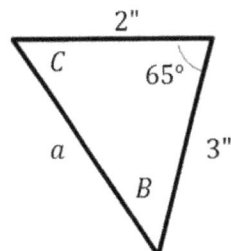

13.9:

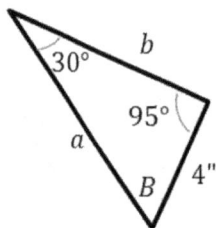

13.10:

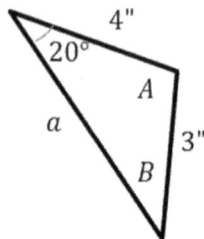

13.11:

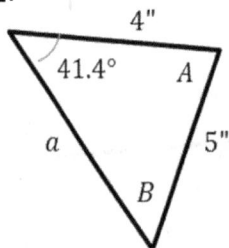

13.12:

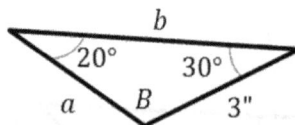

13.13:

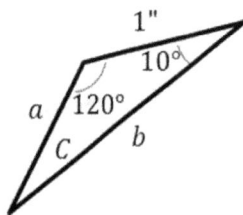

13.14:

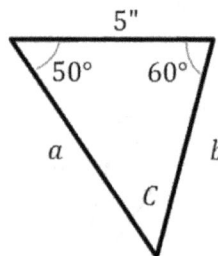

13.15:

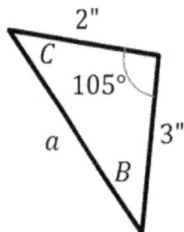

13.16:

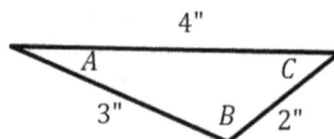

Chapter 14

Show the following:

14.1: $\sinh(-\theta) = -\sinh\theta$
using exponentials

14.2: $\cosh(-\theta) = \cosh\theta$
using exponentials

14.3: $\cosh\theta + \sinh\theta = e^{\theta}$

14.4: $\cosh\theta - \sinh\theta = e^{-\theta}$

14.5: $\sinh(A + B)$
$= \sinh A \cosh B + \sinh B \cosh A$
Hint: see proof of $\sin(A + B)$

14.6: $\cosh(A + B)$
$= \cosh A \cosh B + \sinh A \sinh B$
Hint: see proof of $\sin(A + B)$

14.7: $\sinh(A - B)$
$= \sinh A \cosh B - \sinh B \cosh A$
Hint: Use Problem 14.5

14.8: $\cosh(A - B)$
$= \cosh A \cosh B - \sinh A \sinh B$
Hint: Use Problem 14.6

14.9: $\sinh 2\theta = 2\sinh\theta\cosh\theta$

14.10: $\cosh 2\theta$
$= \cosh^2\theta + \sinh^2\theta$
$= 1 + 2\sinh^2\theta = 2\cosh^2\theta - 1$

14.11: $\cosh\frac{\theta}{2} = \pm\sqrt{\frac{\cosh\theta+1}{2}}$
Hint: Use Problem 14.10

14.12: $\sinh\frac{\theta}{2} = \pm\sqrt{\frac{\cosh\theta-1}{2}}$
Hint: Use Problem 14.10

14.13: $\sinh A \cosh B$
$= \frac{1}{2}[\sinh(A + B) + \sinh(A - B)]$
Hint: Use exponentials

14.14: $\cosh A \cosh B$
$= \frac{1}{2}[\cosh(A + B) + \cosh(A - B)]$
Hint: Use exponentials

14.15: $2\sinh\frac{A+B}{2}\cosh\frac{A-B}{2}$
$= \sinh A + \sinh B$
Hint: Use Problem 14.13

14.16: $2\cosh\frac{A+B}{2}\cosh\frac{A-B}{2}$
$= \cosh A + \cosh B$
Hint: Use Problem 14.14

Appendix C: Solutions to Problem Sets

Chapter 1

1.1: $5 + (-9) - (-4)$ $= 5 - 9 + 4$ $= -4 + 4$ $= 0$	**1.2:** $2 + 8 - (-3)$ $= 2 + 8 + 3$ $= 10 + 3$ $= 13$
1.3: -2×8 $= -16$	**1.4:** $-3 \times -4 \times -2$ $= +12 \times -2$ $= -24$
1.5: $\dfrac{5}{12} - \dfrac{4}{20}$ $\dfrac{5}{12} - \dfrac{4}{20} = \dfrac{25}{60} - \dfrac{12}{60} = \dfrac{13}{60}$	**1.6:** $\dfrac{2}{3} - \dfrac{3}{4} + \dfrac{1}{6}$ $\dfrac{2}{3} - \dfrac{3}{4} + \dfrac{1}{6} = \dfrac{8}{12} - \dfrac{9}{12} + \dfrac{2}{12}$ $= \dfrac{-1}{12} + \dfrac{2}{12} = \dfrac{1}{12}$
1.7: $-\dfrac{3}{2} \times \dfrac{2}{5}$ $-\dfrac{3}{2} \times \dfrac{2}{5} = -\dfrac{6}{10} = -\dfrac{3}{5}$	**1.8:** $\dfrac{3}{4} \times \dfrac{1}{12} \times \dfrac{2}{3}$ $\dfrac{3}{4} \times \dfrac{1}{12} \times \dfrac{2}{3} = \left(\dfrac{3}{48}\right) \times \dfrac{2}{3} = \dfrac{2}{48} = \dfrac{1}{24}$ In second step, the 3's cancel
1.9: $5 \times (2 - 3)$ $= 5 \times (-1) = -5$	**1.10:** $3 \times \left(4 - (8 + 2)\right)$ $= 3(4 - 10) = 3(-6) = -18$
1.11: $(2^2 + 3)^2$ $(4 + 3)^2 = 7^2 = 49$	**1.12:** $\left(\dfrac{18}{(4-2)}\right)\left(\dfrac{(-2)^2}{6}\right)$ $= \left(\dfrac{18}{2}\right)\left(\dfrac{4}{6}\right) = \dfrac{3}{2}\left(\dfrac{4}{1}\right) = 6$

1.13: $x^2 = 2 - x$

$$x^2 + x - 2 = 0$$

1,1   2,1

$1(2) - 1(1)$

$$(x + 2)(x - 1) = 0$$
$$x = \{-2, 1\}$$

1.14: $-x^2 = 2x - 3$

$$x^2 + 2x - 3 = 0$$

1,1   3,1

$1(3) - 1(1)$

$$(x + 3)(x - 1) = 0$$
$$x = \{-3, 1\}$$

1.15: $x^2 = 5 - 4x$

$$x^2 + 4x - 5 = 0$$

1,1   5,1

$1(5) - 1(1)$

$$(x + 5)(x - 1) = 0$$
$$x = \{-5, 1\}$$

1.16: $x^2 + \frac{7x}{2} = 2$

$$2x^2 + 7x - 4 = 0$$

1,2   2,2; 4,1

$2(4) - 1(1)$

$$(x + 4)(2x - 1) = 0$$
$$x = \left\{-4, \frac{1}{2}\right\}$$

Chapter 2

2.1: i^{12} $$i^{12} = (i^4)^3 = 1^3 = 1$$	**2.2:** $(-i)^4$ $$(-i)^4 = (-1)^4(i^4) = 1 \times 1 = 1$$
2.3: $\frac{2i}{i^3}$ $$\frac{2i}{i^3} = \frac{2}{i^2} = \frac{2}{-1} = -2$$	**2.4:** $-4i^3 \times i^2$ $$-4i^3 \times i^2 = -4i^5 = -4(1)i = -4i$$
2.5: $(i+1) - (2i-1)$ $$(i+1) - (2i-1)$$ $$= (1+1) + (i-2i) = 2 - i$$	**2.6:** $\frac{i+1}{i-1}$ $$\frac{i+1}{i-1} = \frac{i+1}{i-1} \times \frac{-i-1}{-i-1}$$ $$= \frac{-i^2-i-i-1}{-i^2-i+i+1} = \frac{1-1-2i}{1+1} = \frac{-2i}{2} = -i$$
2.7: $\frac{(1-i)(1+3i)}{3-2i}$ $$\frac{(1-i)(1+3i)}{3-2i} = \frac{1+3i-i-3i^2}{3-2i} = \frac{1+3+2i}{3-2i}$$ $$= \frac{4+2i}{3-2i} \times \frac{3+2i}{3+2i} = \frac{12+8i+6i+4i^2}{9+6i-6i-4i^2}$$ $$= \frac{12-4+14i}{9+4} = \frac{8}{13} + \frac{14}{13}i$$	**2.8:** $2i - \frac{3+3i}{2+2i}$ $$2i - \frac{3+3i}{2+2i} \times \frac{2-2i}{2-2i} = 2i - \left[\frac{6-6i+6i-6i^2}{4-4i+4i-4i^2}\right]$$ $$= 2i - \left[\frac{6+6}{4+4}\right] = 2i - \frac{3}{2}$$
2.9: $x^2 + 4x + 6$ $$x = \frac{-4 \pm \sqrt{16-4\times6}}{2} = -2 \pm \frac{\sqrt{-8}}{2}$$ $$= -2 \pm \frac{2i\sqrt{2}}{2} = -2 \pm i\sqrt{2}$$	**2.10:** $x^2 + 2x + 4$ $$x = \frac{-2 \pm \sqrt{4-4\times4}}{2} = -1 \pm \frac{\sqrt{-12}}{2}$$ $$= -1 \pm \frac{2i\sqrt{3}}{2} = -1 \pm i\sqrt{3}$$

2.11: $x^2 - x + 9$

$$x = \frac{1 \pm \sqrt{1 - 4 \times 9}}{2} = \frac{1 \pm \sqrt{-35}}{2}$$

$$= \frac{1}{2} \pm \frac{i\sqrt{35}}{2}$$

2.12: $2x^2 - 3x + 2$

$$x = \frac{3 \pm \sqrt{9 - 4 \times 2 \times 2}}{4} = \frac{3 \pm \sqrt{-7}}{4}$$

$$= \frac{3}{4} \pm \frac{i\sqrt{7}}{4}$$

2.13: $(2 + i)$, $(2 - i)$

$$x = 2 + i \rightarrow x - 2 - i = 0;$$
$$x = 2 - i \rightarrow x - 2 + i = 0$$

$$(x + (-2 - i))(x + (-2 + i))$$

$$= x^2 + (-2 + i)x + (-2 - i)x$$
$$+ (4 - 2i + 2i - i^2)$$

$$= x^2 - 2x - 2x + ix - ix + 4 + 1$$

$$= x^2 - 4x + 5$$

2.14: $(3 + 2i)$, $(3 - 2i)$

$$x = 3 + 2i \rightarrow x - 3 - 2i = 0;$$
$$x = 3 - 2i \rightarrow x - 3 + 2i = 0$$

$$(x + (-3 - 2i))(x + (-3 + 2i))$$

$$= x^2 + (-3 + 2i)x + (-3 - 2i)x$$
$$+ (9 - 6i + 6i - 4i^2)$$

$$= x^2 - 3x - 3x + 2ix - 2ix + 9 + 4$$

$$= x^2 - 6x + 13$$

2.15: $(1 + 4i)$, $(1 - 4i)$

$$x = 1 + 4i \rightarrow x - 1 - 4i = 0;$$
$$x = 1 - 4i \rightarrow x - 1 + 4i = 0$$

$$(x + (-1 - 4i))(x + (-1 + 4i))$$

$$= x^2 + (-1 + 4i)x + (-1 - 4i)x$$
$$+ (1 - 4i + 4i - 16i^2)$$

$$= x^2 - x - x + 4ix - 4ix + 1 + 16$$

$$= x^2 - 2x + 17$$

2.16: $(-2 + 4i)$, $(-2 - 4i)$

$$x = -2 + 4i \rightarrow x + 2 - 4i = 0;$$
$$x = -2 - 4i \rightarrow x + 2 + 4i = 0$$

$$(x + (2 - 4i))(x + (2 + 4i))$$

$$= x^2 + (2 + 4i)x + (2 - 4i)x$$
$$+ (4 + 8i - 8i - 16i^2)$$

$$= x^2 + 2x + 2x + 4ix - 4ix + 4 + 16$$

$$= x^2 + 4x + 20$$

Chapter 3

<table>
<tr>
<td>

3.1: $4x - 2y = 6$

Solving for y:

$$y = -\frac{1}{2}(6 - 4x) = 2x - 3$$

There is only one y for every x, so this is a function.

</td>
<td>

3.2: $y = x^2$

This has been solved for y, and there is only one y for every x, so this is a function.

</td>
</tr>
<tr>
<td>

3.3: $x = y^2$

Solving for y:

$$y = \pm\sqrt{x}$$

There is more than one y for every x, so this is not a function.

</td>
<td>

3.4: $y = \begin{cases} x & x \in [0, \infty) \\ -1 & x \in (-\infty, 0] \end{cases}$

At $x = 0, y = 0$ and $y = -1$, so this is not a function. If the top and or bottom expression had open parentheses on the 0, then it would have been a function.

</td>
</tr>
<tr>
<td>

3.5: $f(x) = 2x;$
$\quad\quad g(x) = 5x + 8$

$(f \circ g)(x) = 2(5x + 8) = 10x + 16$

$(g \circ f)(x) = 5(2x) + 8 = 10x + 8$

</td>
<td>

3.6: $f(x) = x + 1;$
$\quad\quad g(x) = 3x - 4$

$(f \circ g)(x) = (3x - 4) + 1 = 3x - 3$

$(g \circ f)(x) = 3(x + 1) - 4 = 3x - 1$

</td>
</tr>
<tr>
<td>

3.7: $f(x) = x + 3;$
$\quad\quad g(x) = 4x - 6$

$(f \circ g)(x) = (4x - 6) + 3 = 4x - 3$

$(g \circ f)(x) = 4(x + 3) - 6 = 4x + 6$

</td>
<td>

3.8: $f(x) = \sqrt{x};$
$\quad\quad g(x) = x^2 + 4x + 4$

$$g(x) = (x + 2)^2$$

$(f \circ g)(x) = \sqrt{(x + 2)^2} = x + 2$

$(g \circ f)(x) = \sqrt{x}^2 + 4\sqrt{x} + 4$
$\quad\quad\quad\quad\quad = x + 4\sqrt{x} + 4$

</td>
</tr>
</table>

3.9: $f(x) = 3x^2$;
  $g(x) = 2x + 4$

  $$(f \circ g)(x) = 3(2x + 4)^2$$
  $$= 3[4x^2 + 16x + 16]$$
  $$= 12x^2 + 48x + 48$$

  $$(g \circ f)(x) = 2(3x^2) + 4 = 6x^2 + 4$$

3.10: $f(x) = x^2 - 2x$;
  $g(x) = x - 3$

  $$(f \circ g)(x) = (x - 3)^2 - 2(x - 3)$$
  $$= x^2 - 6x + 9 - 2x + 6$$
  $$= x^2 - 8x + 15$$

  $$(g \circ f)(x) = x^2 - 2x - 3$$

3.11: $f(x) = x^2 - 3x + 9$;
  $g(x) = x + 2$

  $$(f \circ g)(x) = (x + 2)^2 - 3(x + 2) + 9$$
  $$= x^2 + 4x + 4 - 3x - 6 + 9$$
  $$= x^2 + x + 7$$

  $$(g \circ f)(x) = x^2 - 3x + 9 + 2$$
  $$= x^2 - 3x + 11$$

3.12: $f(x) = (3 + 2i)x$;
  $g(x) = \Re\{4x\}$

  $$(f \circ g)(x) = (3 + 2i)\Re\{4x\}$$
  $$= (3 + 2i)(4x) = (12 + 8i)x$$

  $$(g \circ f)(x) = \Re\{4(3 + 2i)x\}$$
  $$= \Re\{(12 + 8i)x\} = 12x$$

3.13: $f(x) = 3x + 4$

  $$y = 3x + 4 \to x = 3y + 4$$

  $$y = \frac{x - 4}{3}$$

  $$f^{-1}(x) = \frac{x - 4}{3}$$

3.14: $f(x) = x^3 + 3$

  $$y = x^3 + 3 \to x = y^3 + 3$$

  $$y = \sqrt[3]{x - 3}$$

  $$f^{-1}(x) = \sqrt[3]{x - 3}$$

3.15: $f(x) = 3\sqrt{x} - 5$

  $$y = 3\sqrt{x} - 5 \to x = 3\sqrt{y} - 5$$

  $$y = \left(\frac{x + 5}{3}\right)^2$$

  $$f^{-1}(x) = \left(\frac{x + 5}{3}\right)^2$$

3.16: $f(x) = (1 - i)x + 2$

  $$y = (1 - i)x + 2 \to x = (1 - i)y + 2$$

  $$y = \frac{x - 2}{1 - i} = \frac{[(x - 2)(1 + i)]}{1 + 1}$$

  $$f^{-1}(x) = \frac{[(x - 2)(1 + i)]}{2}$$

Chapter 4

4.1: $\log_5 2 = x$ $\qquad 2 = 5^x$	**4.2:** $\log_3 x = 8$ $\qquad x = 3^8$
4.3: $\log x = 20$ $\qquad x = 10^{20}$	**4.4:** $\ln 4 = x$ $\qquad 4 = e^x$
4.5: $25 = \dfrac{1}{5^x}$ Try finding like bases: $\qquad 5^2 = 5^{-x}$ $\qquad 2 = -x; \ x = -2$	**4.6:** $16^x = 4$ Try finding like bases: $\qquad (4^2)^x = 4^1$ $\qquad 2x = 1; \ x = 1/2$
4.7: $3^{(x^2+2x)} = 27$ Try finding like bases: $\qquad 3^{(x^2+2x)} = 3^3$ $\qquad x^2 + 2x - 3 = 0$ $\qquad (x+3)(x-1) = 0$ $\qquad x = \{-3,1\}$	**4.8:** $32^x(2^{3x}) = 16$ Try finding like bases: $\qquad (2^5)^x (2^{3x}) = 2^4$ $\qquad 2^{5x}(2^{3x}) = 2^4$ $\qquad 2^{8x} = 2^4$ $\qquad 8x = 4; \ x = 1/2$
4.9: $5^{2x} = 6$ There are no like bases, so take the ln of both sides: $\qquad \ln 5^{2x} = \ln 6$ $\qquad 2x \ln 5 = \ln 6$ $\qquad x = \dfrac{\ln 6}{2 \ln 5} = 0.517$	**4.10:** $4^{(x+1)} = 3^x$ There are no like bases, so take the ln of both sides: $\qquad \ln 4^{(x+1)} = \ln 3^x$ $\qquad (x+1) \ln 4 = x \ln 3$ $\qquad x(\ln 4 - \ln 3) = -\ln 4$ $\qquad x = -\dfrac{\ln 4}{\ln 4/3} = -4.819$

4.11: $5^{(x-3)} = \dfrac{1}{2^x}$ There are no like bases, so take the ln of both sides: $$\ln 5^{(x-3)} = \ln 2^{-x}$$ $$(x-3)\ln 5 = -x\ln 2$$ $$x(\ln 5 + \ln 2) = 3\ln 5$$ $$x = \frac{3\ln 5}{\ln 10} = 2.097$$	**4.12:** $3^{2x} = \dfrac{1}{16^{(x+1)}}$ There are no like bases, so take the ln of both sides: $$\ln 3^{2x} = \ln 16^{-(x+1)}$$ $$2x\ln 3 = -(x+1)\ln 16$$ $$x(2\ln 3 + \ln 16) = -\ln 16$$ $$x\ln(3^2 \times 16) = -\ln 16$$ $$x = -\frac{\ln 16}{\ln 144} = -0.558$$
4.13: $\log_2 40$ $$\log_2 40 = \frac{\ln 40}{\ln 2} = 5.322$$	**4.14:** $\log_5 100$ $$\log_5 100 = \frac{\ln 100}{\ln 5} = 2.861$$
4.15: $\log_7 8$ $$\log_7 8 = \frac{\ln 8}{\ln 7} = 1.069$$	**4.16:** $\log_{1/4} 3$ $$\log_{1/4} 3 = \frac{\ln 3}{\ln 1/4} = -0.792$$

Chapter 5

5.1: $4y + 8x = 16$ Plot $y = 4 - 2x$: 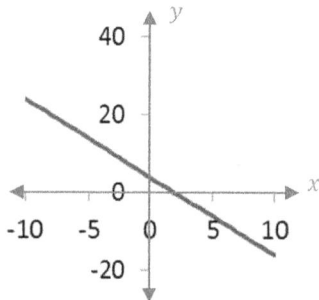	**5.2:** $2y - 6 = 12x$ Plot $y = 6x - 3$: 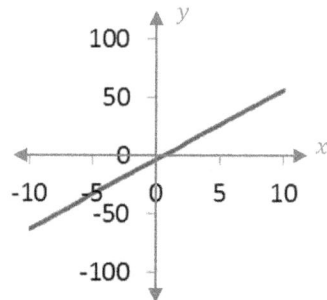
5.3: $5y = 15\sqrt{x} + 40$ Plot $y = 3\sqrt{x} + 8$: 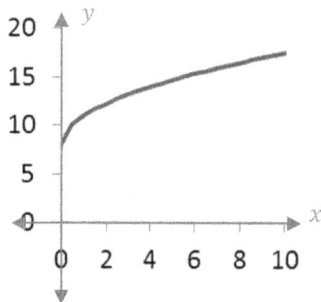	**5.4:** $3y = x^2 - 9$ Plot $y = \dfrac{x^2}{3} - 3$: 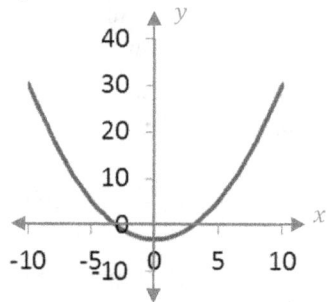
5.5: $y = 3x^3 - 2x$ 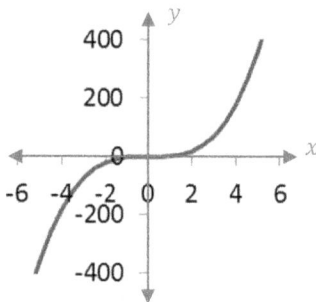	**5.6:** $y = x^2 - 5x^3$ 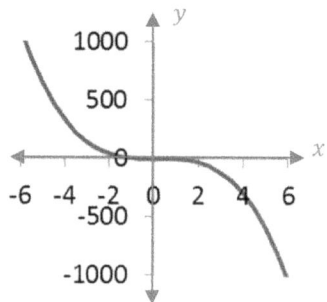

5.7: $y = x - 4x^3$

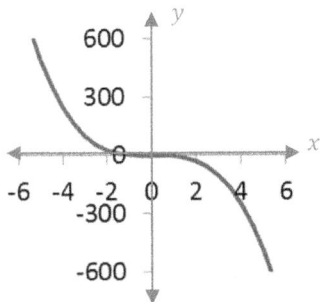

5.8: $y = x^3 - 3x^2 - 4x - 5$

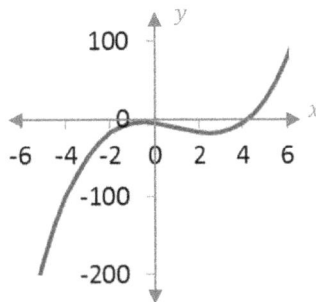

5.9: $y = \log(10x + 2)$

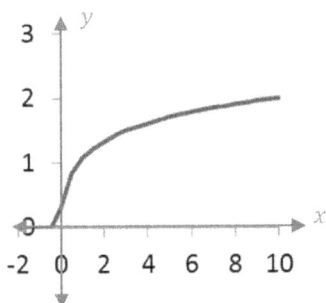

5.10: $y = \ln(x^2 + 2)$

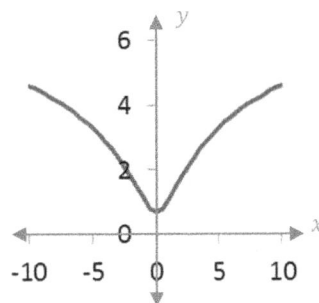

5.11: $y = x^2 \ln x$

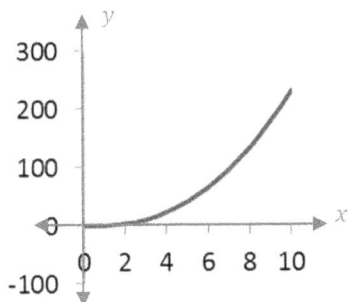

5.12: $y = \log_2 x$

Plot $y = \dfrac{\ln x}{\ln 2}$:

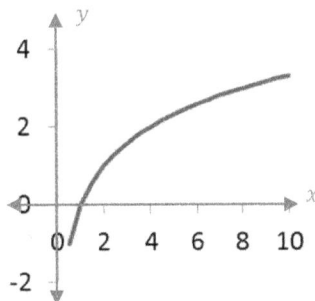

5.13: $y = 3^x + x^3$

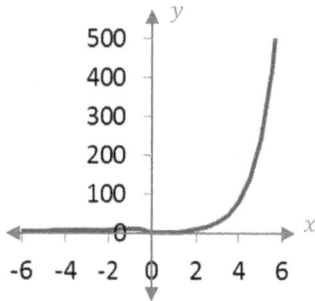

5.14: $y = 4^{(x+1)}$

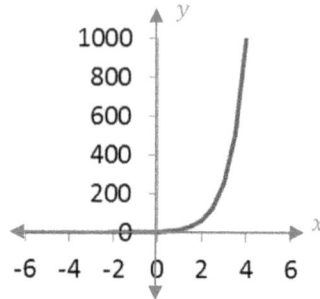

5.15: $y = x^x$

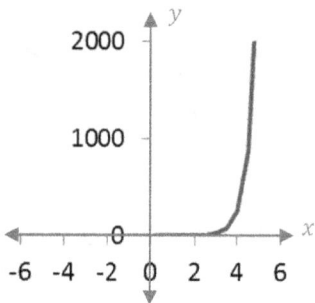

5.16: $y = 2^{-x} + 2^{x^2}$

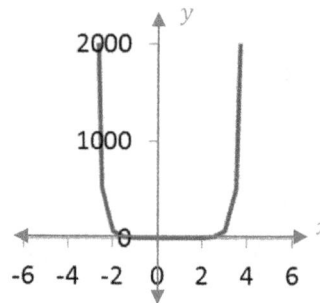

For these chapter problems, graphing software such as Excel was used.

Chapter 6

6.1: $(1,2); (3,4)$	**6.2:** $(-5,2); (3,8)$
$d = \sqrt{(3-1)^2+(4-2)^2} = \sqrt{8} = 2\sqrt{2}$	$d = \sqrt{(3+5)^2+(8-2)^2} = 10$
$midpoint = \left(\frac{3+1}{2}, \frac{4+2}{2}\right) = (2,3)$	$midpoint = \left(\frac{3-5}{2}, \frac{8+2}{2}\right) = (-1,5)$
$m = \frac{4-2}{3-1} = 1$	$m = \frac{8-2}{3+5} = \frac{6}{8} = \frac{3}{4}$
6.3: $(-1,10); (-2,6)$	**6.4:** $(1,0); (-1,2)$
$d = \sqrt{(-2+1)^2+(6-10)^2} = \sqrt{17}$	$d = \sqrt{(-1-1)^2+(2-0)^2} = 2\sqrt{2}$
$midpoint = \left(\frac{-2-1}{2}, \frac{6+10}{2}\right) = \left(-\frac{3}{2}, 8\right)$	$midpoint = \left(\frac{-1+1}{2}, \frac{2+0}{2}\right) = (0,1)$
$m = \frac{6-10}{-2+1} = \frac{-4}{-1} = 4$	$m = \frac{2-0}{-1-1} = \frac{2}{-2} = -1$
6.5: $(-5,-5); (2,2)$	**6.6:** $(4,-3); (8-1)$
$d = \sqrt{(2+5)^2+(2+5)^2} = 7\sqrt{2}$	$d = \sqrt{(8-4)^2+(-1+3)^2} = 2\sqrt{5}$
$midpoint = \left(\frac{2-5}{2}, \frac{2-5}{2}\right) = \left(-\frac{3}{2}, -\frac{3}{2}\right)$	$midpoint = \left(\frac{8+4}{2}, \frac{-1-3}{2}\right) = (6,-2)$
$m = \frac{2+5}{2+5} = 1$	$m = \frac{-1+3}{8-4} = \frac{2}{4} = \frac{1}{2}$
6.7: For Problem 6.1: $(1,2); (3,4)$	**6.8:** For Problem 6.2: $(-5,2); (3,8)$
Plug into point-slope form with $(a,b) = (1,2)$ and $m = 1$, then reduce:	Use $(a,b) = (-5,2)$ and $m = \frac{3}{4}$, then reduce:
$y - 2 = 1(x-1)$	$y - 2 = \frac{3}{4}(x+5)$
$y = x + 1$	$y = \frac{3}{4}x + \frac{23}{4}$
The y-intercept is $(0,1)$.	The y-intercept is $(0, 23/4)$.

6.9: For Problem 6.3:
$$(-1,10); (-2,6)$$

Use $(a, b) = (-1,10)$ and $m = 4$, then reduce:

$$y - 10 = 4(x + 1)$$

$$y = 4x + 14$$

The y-intercept is $(0,14)$.

6.10: For Problem 6.4:
$$(1,0); (-1,2)$$

Use $(a, b) = (1,0)$ and $m = -1$, then reduce:

$$y - 0 = -1(x - 1)$$

$$y = -x + 1$$

The y-intercept is $(0,1)$.

6.11: $2y + 4x = 6$

Put the equation into the proper form:

$$y = -2x + 3$$

The slope is -2; the y-intercept is $(0,3)$.

6.12: $2(3 - y) = 6x$

Put the equation into the proper form:

$$y = -3x + 3$$

The slope is -3; the y-intercept is $(0,3)$.

6.13: $4 - y = 5x + 2$

Put the equation into the proper form:

$$y = -5x + 2$$

The slope is -5; the y-intercept is $(0,2)$.

6.14: $x + 5 = 9 - y$

Put the equation into the proper form:

$$y = -x + 4$$

The slope is -1; the y-intercept is $(0,4)$.

6.15: $y = 3x - 1$; $(2,1)$

The slope remains the same: $m = 3$.
In point-slope form:

$$(y - 1) = 3(x - 2)$$

Which reduces to :

$$y = 3x - 5$$

6.16: $y = 3x - 1$; $(2,1)$

The slope becomes: $m = -1/3$.
In point-slope form:

$$(y - 1) = -\frac{1}{3}(x - 2)$$

Which reduces to :

$$y = \frac{-x+5}{3}$$

Chapter 7

7.1: A circle of radius 4 centered on the origin.	**7.2:** A circle of diameter 4 centered on the point $(-1,5)$.
With $r = 4$ and $(h,k) = (0,0)$, we have: $$(x-h)^2+(y-k)^2 = r^2$$ $$x^2 + y^2 = 16$$	With $r = \frac{4}{2} = 2$ and $(h,k) = (-1,5)$: $$(x-h)^2+(y-k)^2 = r^2$$ $$(x+1)^2+(y-5)^2 = 4$$ $$x^2 + y^2 + 2x - 10y + 26 = 16$$ $$x^2 + y^2 + 2x - 10y = -10$$
7.3: A parabola that opens to the right with $|p| = 3$.	**7.4:** A parabola that opens downward with $|p| = 2$.
Opening to the right means we have a focus at $(3,0)$ and a directrix of $x = -3$. The equation is: $$y^2 = 4px = 12x$$	Opening downward means we have a focus at $(0,-2)$ and a directrix of $y = 2$. The equation is: $$x^2 = 4py = -8y$$
7.5: An ellipse with major and minor axes endpoints at $(\pm4,0), (0,\pm2)$.	**7.6:** An ellipse with a major axis from $(0,\pm5)$ and foci at $(0,\pm4)$.
Since $4 > 2, a = 4, b = 2$, and the major axis is along the x direction: $$\frac{x^2}{a^2} + \frac{y^2}{b^2} = 1$$ $$\frac{x^2}{16} + \frac{y^2}{4} = 1$$ $$x^2 + 4y^2 = 16$$	We know $a = 5, c = 4$, which gives us: $$b = \sqrt{a^2 - c^2} = \sqrt{25 - 16} = \sqrt{9} = 3$$ With the major axis along y, we get: $$\frac{x^2}{b^2} + \frac{y^2}{a^2} = 1$$ $$\frac{x^2}{9} + \frac{y^2}{25} = 1$$

7.7: A hyperbola with foci at $(\pm\sqrt{13}, 0)$ and vertices at $(\pm3, 0)$.

We know $a = 3, c = \sqrt{13}$, so:

$$b = \sqrt{c^2 - a^2} = \sqrt{13 - 9} = \sqrt{4} = 2$$

With the major axis along x, we get:

$$\frac{x^2}{a^2} - \frac{y^2}{b^2} = 1$$

$$\frac{x^2}{9} - \frac{y^2}{4} = 1$$

$$4x^2 - 9y^2 = 36$$

7.8: A hyperbola with vertices at $(0, \pm12)$ and asymptotes of $y = \pm6x$.

We know $a = 12$ and the major axis is along y. The asymptotes are:

$$y = \pm\frac{ax}{b} = \pm\frac{12x}{b} = \pm6x \rightarrow b = 2$$

Therefore:

$$\frac{y^2}{a^2} - \frac{x^2}{b^2} = 1$$

$$\frac{y^2}{144} - \frac{x^2}{4} = 1$$

$$y^2 - 36x^2 = 144$$

7.9: $(x - 1)^2 + (y + 1)^2 = 4$

We have the form:

$$(x - h)^2 + (y - k)^2 = r^2$$

So, $r = 2$, and $(h, k) = (1, -1)$.

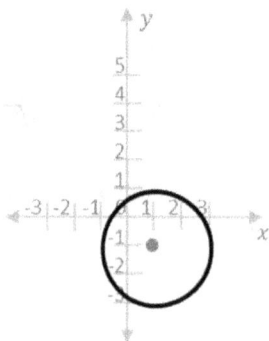

7.10: $(x + 2)^2 + (y - 3)^2 = 1$

We have the form:

$$(x - h)^2 + (y - k)^2 = r^2$$

So, $r = 1$, and $(h, k) = (-2, 3)$.

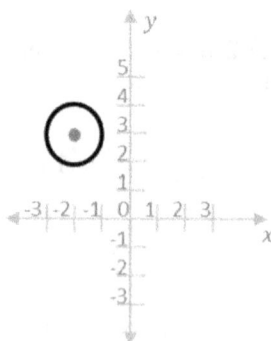

7.11: $x^2 = 16y$

We have the form:
$$x^2 = 4py$$

So, $p = 4$, the focus is $(0,4)$, and the directrix is $y = -4$. To help sketch, we note that at $x = \pm 4, y = 1$.

7.12: $y^2 = -8x$

We have the form:
$$y^2 = 4px$$

So, $p = -2$, the focus is $(-2,0)$, and the directrix is $x = 2$. To help sketch, we note that at $y = \pm 4, x = -2$.

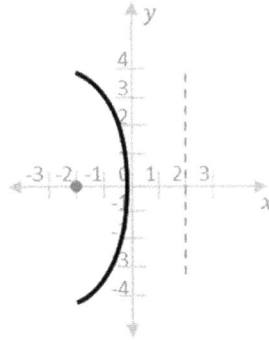

7.13: $\dfrac{x^2}{16} + \dfrac{y^2}{4} = 1$

We have the form:
$$\frac{x^2}{a^2} + \frac{y^2}{b^2} = 1$$

So the vertices are at $(\pm 4, 0)$ and $(0, \pm 2)$.

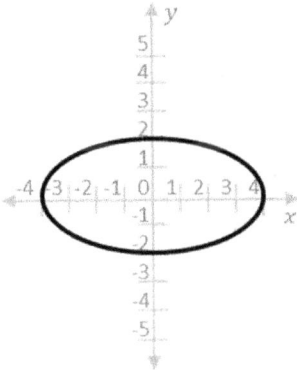

7.14: $\dfrac{y^2}{25} + \dfrac{x^2}{9} = 1$

We have the form:
$$\frac{y^2}{a^2} + \frac{x^2}{b^2} = 1$$

So the vertices are at $(\pm 3, 0)$ and $(0, \pm 5)$.

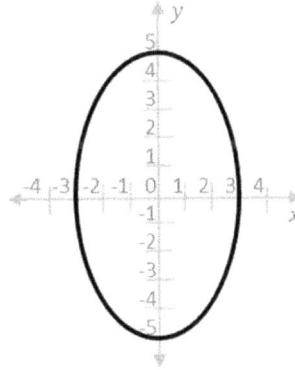

7.15: $\frac{x^2}{9} - \frac{y^2}{4} = 1$

We have the form:
$$\frac{x^2}{a^2} - \frac{y^2}{b^2} = 1$$

So the vertices are at $(\pm 3, 0)$ and the asymptotes are given by

$$y = \frac{\pm bx}{a} = \frac{\pm 2x}{3}.$$

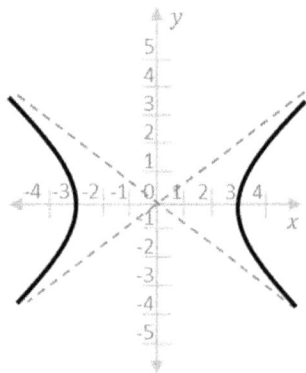

7.16: $\frac{y^2}{16} - \frac{x^2}{4} = 1$

We have the form:
$$\frac{y^2}{a^2} - \frac{x^2}{b^2} = 1$$

So the vertices are at $(0, \pm 4)$ and the asymptotes are given by

$$y = \frac{\pm ax}{b} = \frac{\pm 4x}{2} = \pm 2x.$$

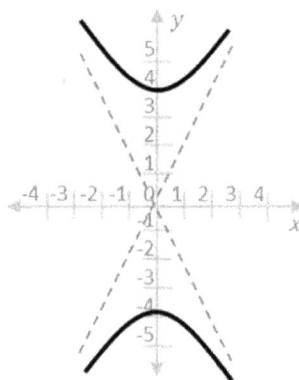

Chapter 8

8.1: $2y = 2x + 3$; $y = 4x$	**8.2:** $y = x - 2$; $2y = 3x + 2$
$-2(y = 4x) \rightarrow -2y = -8x$	$-2(y = x - 2) \rightarrow -2y = -2x + 4$
$\begin{aligned} 2y &= 2x + 3 \\ + \quad -2y &= -8x \\ \hline 0 &= -6x + 3 \end{aligned}$	$\begin{aligned} 2y &= 3x + 2 \\ + \quad -2y &= -2x + 4 \\ \hline 0 &= x + 6 \end{aligned}$
$x = \dfrac{3}{6} = \dfrac{1}{2}$	$x = -6$
$y = 4x = 2$	$y = x - 2 = -6 - 2 = -8$
Solution is $\left(\dfrac{1}{2}, 2\right)$	Solution is $(-6, -8)$
8.3: $3y = x^2 + 8$; $y = 4 - x$	**8.4:** $4y = x^2 + 7$; $y = x + 1$
$-3(y = 4 - x) \rightarrow -3y = -12 + 3x$	$-4(y = x + 1) \rightarrow -4y = -4x - 4$
$\begin{aligned} 3y &= x^2 \qquad + 8 \\ + \quad -3y &= \qquad 3x - 12 \\ \hline 0 &= x^2 + 3x - 4 \end{aligned}$	$\begin{aligned} 4y &= x^2 \qquad + 7 \\ + \quad -4y &= \qquad -4x - 4 \\ \hline 0 &= x^2 - 4x + 3 \end{aligned}$
$(x + 4)(x - 1) = 0$ $x = \{-4, 1\}$	$(x - 3)(x - 1) = 0$ $x = \{3, 1\}$
$y = 4 - x = 4 + 4 = 8;$ $y = 4 - x = 4 - 1 = 3;$	$y = x + 1 = 3 + 1 = 4;$ $y = x + 1 = 1 + 1 = 2;$
Solutions are $(-4, 8), (1, 3)$	Solutions are $(3, 4), (1, 2)$

8.5: $y = x + 3;\ 3y = 2x^2 + 10$

$$-3(y = x + 3) \rightarrow -3y = -3x - 9$$

$$\begin{array}{l} 3y = 2x^2 \quad\ + 10 \\ + \ -3y = \quad\ -3x - 9 \\ \hline 0 = 2x^2 - 3x + 1 \end{array}$$

$$(2x - 1)(x - 1) = 0$$
$$x = \{\tfrac{1}{2}, 1\}$$

$$y = x + 3 = 3 + \tfrac{1}{2} = \tfrac{7}{2};$$
$$y = x + 3 = 1 + 3 = 4;$$

Solutions are $\left(\tfrac{1}{2}, \tfrac{7}{2}\right), (1,4)$

8.6: $y = -x;\ 2y = x^2 - 3$

$$-2(y = -x) \rightarrow -2y = 2x$$

$$\begin{array}{l} 2y = x^2 \quad\ - 3 \\ + \ -2y = \quad\ 2x \\ \hline 0 = x^2 + 2x - 3 \end{array}$$

$$(x + 3)(x - 1) = 0$$
$$x = \{-3, 1\}$$

$$y = -x = 3;$$
$$y = -x = -1;$$

Solutions are $(-3, 3), (1, -1)$

8.7: $y = x + 1;\ 4y = 5x - 1$

Plug 1st equation into 2nd:

$$4(x + 1) = 5x - 1$$
$$4x + 4 = 5x - 1$$
$$-x = -5$$
$$x = 5$$

$$y = x + 1 = 5 + 1 = 6$$

Solution is $(5,6)$

8.8: $y = 3x + 3;\ 2y = x + 10$

Plug 1st equation into 2nd:

$$2(3x + 3) = x + 10$$
$$6x + 6 = x + 10$$
$$5x = 4$$
$$x = \tfrac{4}{5}$$

$$y = 3x + 3 = \tfrac{12}{5} + \tfrac{15}{5} = \tfrac{27}{5}$$

Solution is $\left(\tfrac{4}{5}, \tfrac{27}{5}\right)$

8.9: $y = 4x - 2$; $3y = 4x^2 + 2$

Plug 1st equation into 2nd:

$$3(4x - 2) = 4x^2 + 2$$
$$12x - 6 = 4x^2 + 2$$
$$4x^2 - 12x + 8 = 0$$
$$(4x - 4)(x - 2) = 0$$
$$x = \{1,2\}$$

$$y = 4x - 2 = 4 - 2 = 2$$
$$y = 4x - 2 = 8 - 2 = 6$$

Solutions are $(1,2), (2,6)$

8.10: $y = 10x + 4$; $\frac{y}{2} = 4x + 3$

Plug 1st equation into 2nd:

$$\tfrac{1}{2}(10x + 4) = 4x + 3$$
$$5x + 2 = 4x + 3$$
$$x = 1$$

$$y = 10x + 4 = 10 + 4 = 14$$

Solution is $(1,14)$

8.11: $y = 2x^2$; $3y = 5x + 1$

Plug 1st equation into 2nd:

$$3(2x^2) = 5x + 1$$
$$6x^2 = 5x + 1$$
$$6x^2 - 5x - 1 = 0$$
$$(x - 1)(6x + 1) = 0$$
$$x = \{1, -\tfrac{1}{6}\}$$

$$y = 2x^2 = 2$$
$$y = 2x^2 = \frac{2}{36} = \frac{1}{18}$$

Solutions are $(1,2), (-\tfrac{1}{6}, \tfrac{1}{18})$

8.12: $y = x - 1$; $2y = x^2 - 1$

Plug 1st equation into 2nd:

$$2(x - 1) = x^2 - 1$$
$$2x - 2 = x^2 - 1$$
$$x^2 - 2x + 1 = 0$$
$$(x - 1)(x - 1) = 0$$
$$x = 1$$

$$y = x - 1 = 1 - 1 = 0$$

Solution is $(1,0)$

8.13: $y = 3x - 2$; $2y = 4x + 5$

Plot $y = 3x - 2$; $y = 2x + \frac{5}{2}$

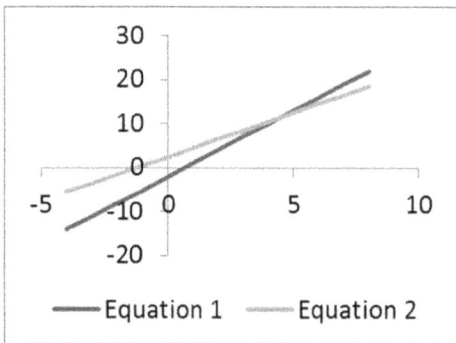

Solution is the crossover at $\left(\frac{9}{2}, \frac{23}{2}\right)$

8.14: $y = x + 2$; $y = x^2$

Plot $y = x + 2$; $y = x^2$

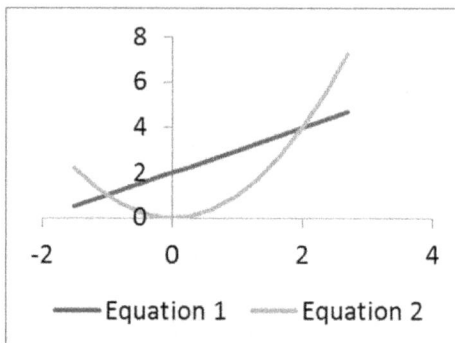

Solutions are the crossovers at
$(-1,1), (2,4)$

8.15: $y = 4x - 1$; $y = x^3 - 3$

Plot $y = 4x - 1$; $y = x^3 - 3$

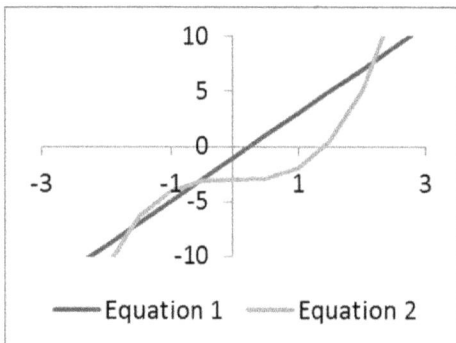

Solutions are the crossovers at
approximately
$(-1.7, -7.7), (-0.5, -3.2), (2.2, 7.8)$

8.16: $y = 3x + 5$; $y = x^2 + 1$

Plot $y = 3x + 5$; $y = x^2 + 1$

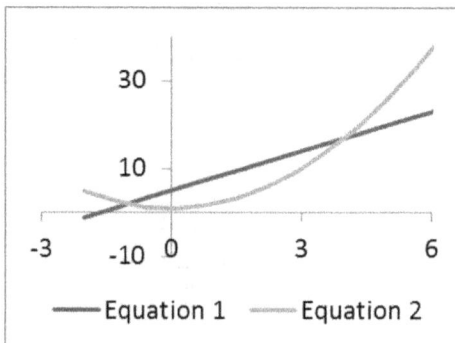

Solutions are the crossovers at
$(-1,2), (4,17)$

Chapter 9

9.1:

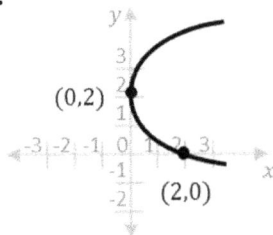

To open along x, we have the following form:
$$(y - k)^2 = 4p(x - h)$$
The vertex is:
$$(h, k) = (0,2) \rightarrow h = 0; k = 2$$

We have the point $(x, y) = (2,0)$, so plug it in:
$$(y - 2)^2 = 4px$$
$$(-2)^2 = 4p(2) \rightarrow p = 1/2$$

The final equation is:
$$(y - 2)^2 = 2x$$

9.2:

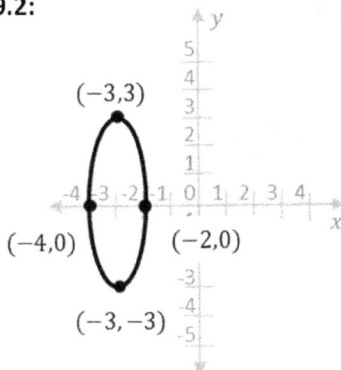

The major axis is on y, so the form is:
$$\frac{(y-k)^2}{a^2} + \frac{(x-h)^2}{b^2} = 1$$

$$(h + b, k) = (-2,0) \rightarrow k = 0$$
$$(h, k + a) = (-3,3) \rightarrow h = -3$$
$$k + a = 3 \rightarrow a = 3$$
$$h + b = -2 \rightarrow b = 1$$

Plugging in:
$$\frac{y^2}{9} + (x + 3)^2 = 1$$

9.3:

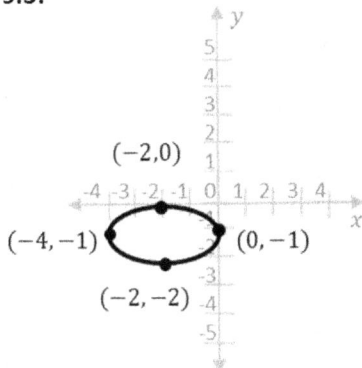

The major axis is on x, so the form is:
$$\frac{(x-h)^2}{a^2} + \frac{(y-k)^2}{b^2} = 1$$

$$(h + a, k) = (0,-1) \rightarrow k = -1$$
$$(h, k + b) = (-2,0) \rightarrow h = -2$$
$$k + b = 0 \rightarrow b = 1$$
$$h + a = 0 \rightarrow a = 2$$

Plugging in:
$$\frac{(x+2)^2}{4} + (y + 1)^2 = 1$$

9.4:

$2y = -x + 4$ $2y = x + 8$

$(-4,3)$ $(0,3)$

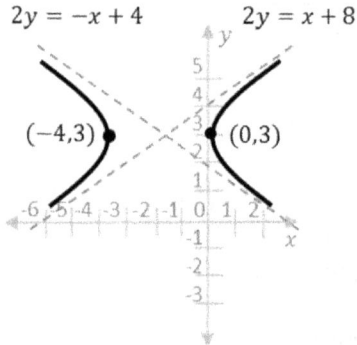

The figure opens along x, so the form is:

$$\frac{(x-h)^2}{a^2} - \frac{(y-k)^2}{b^2} = 1$$

$$(h + a, k) = (0,3) \rightarrow k = 3; h = -a$$

The positively sloped asymptote in $y = mx + b$ form is:

$$y = \frac{b(x-h)}{a} + k$$

$$y = \frac{bx}{a} + \left(k - \frac{bh}{a}\right)$$

We have: $2y = x + 8 \rightarrow y = \frac{x}{2} + 4$

The slope is :

$$\frac{b}{a} = \frac{1}{2} \rightarrow b = a/2$$

The y-intercept is:

$$4 = \left(k - \frac{bh}{a}\right) = 3 - \frac{a}{2}\left(-\frac{a}{a}\right) = 3 + \frac{a}{2} \rightarrow a = 2$$

$$b = \frac{a}{2} = 1; \ \ h = -a = -2$$

Plugging it all in:

$$\frac{(x+2)^2}{4} - (y - 3)^2 = 1$$

Original in black; Reflected in dark grey; Rotated in light grey.	**9.5:** Reflect $y = 2x + 1$ about the x-axis Plug in $-y$ for y: $$-y = 2x + 1$$ $$y = -2x - 1$$ We plot slope -2 and intercept $(-1, 0)$.

9.6: Rotate $y = 2x + 1$ by $90°$

$$x' = x \cos \theta + y \sin \theta = x \cdot 0 + y \cdot 1 = y$$

$$y' = -x \sin \theta + y \cos \theta = -x \cdot 1 + y \cdot 0 = -x$$

$$-x = 2y + 1 \rightarrow y = -\frac{x}{2} - \frac{1}{2}$$

We plot slope $-\frac{1}{2}$ and intercept $(-\frac{1}{2}, 0)$.

Original in black $((x + 2)^2 + (y - 3)^2 = 1)$; Translated in dark grey; Reflected in light grey.	**9.7:** Translate 4 units down and 1 unit right With $h = 1; k = -4$ we plug in $x - 1; y + 4$: $$((x - 1) + 2)^2 + ((y + 4) - 3)^2 = 1$$ $$(x + 1)^2 + (y + 1)^2 = 1$$ This is a circle with radius 1 centered at $(-1, -1)$.

9.8: Reflect about the y-axis

Plug in $-x$ for x:

$$(-x + 2)^2 + (y - 3)^2 = 1$$

Pulling out $(-1)^2 = 1$ from the x term:

$$(x - 2)^2 + (y - 3)^2 = 1$$

This is a circle with radius 1 centered at $(2, 3)$.

9.9: Translate $y = x^2$ 2 units up and 1 unit left

With $h = -1; k = 2$ we plug in $x + 1; y - 2$:

$$(x + 1)^2 = y - 2$$

The new vertex is at $(h, k) = (-1,2)$. Picking $x = 0 \rightarrow$ $y = 3$ and $x = -2 \rightarrow y = 3$ helps us sketch.

9.10: Rotate $y = x^2$ by 90°

$$x' = x \cos \theta + y \sin \theta = x \cdot 0 + y \cdot 1 = y$$
$$y' = -x \sin \theta + y \cos \theta = -x \cdot 1 + y \cdot 0 = -x$$

The equation becomes: $-x = y^2$

The new vertex is at $(h, k) = (0,0)$. Picking $x = -1 \rightarrow$ $y = \pm 1$ and $x = -4 \rightarrow y = \pm 2$ helps.

Original in black;
Translated in dark grey;
Rotated in light grey.

9.11: Reflect $\frac{x^2}{4} + (y - 1)^2 = 1$ about the x-axis

Plug in $-y$: $\frac{x^2}{4} + (-y - 1)^2 = 1 \rightarrow \frac{x^2}{4} + (y + 1)^2 = 1$

Now $h = 0; k = -1; a = 2; b = 1$ and vertices are $(-2, -1), ((0,0), (2, -1), (0, -2)$.

9.12: Rotate $\frac{x^2}{4} + (y - 1)^2 = 1$ by 60°

$$x' = \frac{1}{2}(x + \sqrt{3}y); \quad y' = \frac{1}{2}(y - \sqrt{3}x)$$

$$\frac{(x+\sqrt{3}y)^2}{16} + \frac{(y-\sqrt{3}x-1)^2}{4} = 1$$

For the axes we solve four systems of equations with $(x', y') = (0,0), (-2,1), (0,2)$, and $(2,1)$. For example:

$$0 = \frac{1}{2}(x + \sqrt{3}y) \text{ and } 0 = \frac{1}{2}(y - \sqrt{3}x)$$

The final axes points are:
$(0,0), (-1.9, -1.2), (-1.7,1), (0.1,2.2)$

Original in black;
Translated in dark grey;
Rotated in light grey.

9.13: Translate 3 units down and 2 units left

With $h = -2; k = -3$ we plug in $x + 2; y + 3$:

$$\frac{((y+3)-3)^2}{4} + ((x + 2) - 1)^2 = 1 \rightarrow \frac{y^2}{4} + (x + 1)^2 = 1$$

Now we have $h = -1; k = 0; a = 2; b = 1$.

Axes are:

$$(h, k \pm a) = (-1, \pm 2); \ (h \pm b, k) = (-2,0), (0,0)$$

9.14: Reflect about the y-axis

Plug in $-y$ for y:

$$\frac{(-y-3)^2}{4} + (x - 1)^2 = 1 \rightarrow \frac{(y+3)^2}{4} + (x - 1)^2 = 1$$

Now we have $h = 1; k = -3; a = 2; b = 1$. Axes are:

$$(h, k \pm a) = (1, -5), (1, -1);$$
$$(h \pm b, k) = (0, -3), (2, -3)$$

Original in black

$(\frac{(y-3)^2}{4} + (x - 1)^2 = 1)$;

Translated in dark grey;
Rotated in light grey.

9.15: Translate $\frac{x^2}{9} - \frac{y^2}{4} = 1$, 2 units up and 1 unit right

With $h = 1; k = 2$ we plug in $x - 1; y - 2$:

$$\frac{(x-1)^2}{9} - \frac{(y-2)^2}{4} = 1, \ a = 3, b = 2$$

Vertices are $(h \pm a, k) = (-2,2), (4,2)$

Asymptotes are: $y = \frac{\pm b(x-h)}{a} + k \rightarrow$

$$y = \frac{bx}{a} + k - \frac{bh}{a} = \frac{2x}{3} + \frac{4}{3};$$

Original in black;
Translated in dark grey.

$$y = -\frac{bx}{a} + k + \frac{bh}{a} = -\frac{2x}{3} + \frac{8}{3}$$

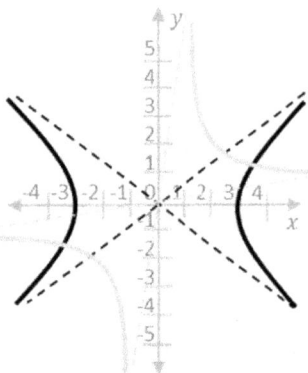

Original in black;
Rotated in light grey.

9.16: Rotate $\dfrac{x^2}{9} - \dfrac{y^2}{4} = 1$ by 45°

$$x' = \frac{\sqrt{2}}{2}(x + y); \quad y' = \frac{\sqrt{2}}{2}(y - x)$$

$$\frac{2}{36}(x + y)^2 - \frac{2}{16}(y - x)^2 = 1$$

The vertex $(-3,0)$ becomes:

$$-3 = \frac{\sqrt{2}}{2}(x + y);$$

$$0 = \frac{\sqrt{2}}{2}(y - x) \rightarrow y = x$$

$$-3 = \frac{\sqrt{2}}{2}(2x) \rightarrow x = y = \frac{-3}{\sqrt{2}}; \ \left(\frac{-3}{\sqrt{2}}, \frac{-3}{\sqrt{2}}\right).$$

Similarly, the vertex $(3,0)$ becomes:

$$3 = \frac{\sqrt{2}}{2}(x + y);$$

$$0 = \frac{\sqrt{2}}{2}(y - x) \rightarrow y = x$$

$$3 = \frac{\sqrt{2}}{2}(2x) \rightarrow x = y = \frac{3}{\sqrt{2}}; \ \left(\frac{3}{\sqrt{2}}, \frac{3}{\sqrt{2}}\right).$$

The asymptote $y = \dfrac{bx}{a} = \dfrac{2x}{3}$ becomes:

$$\frac{\sqrt{2}}{2}(y - x) = \frac{2}{3} \cdot \frac{\sqrt{2}}{2}(y + x) \rightarrow y = 5x$$

The asymptote $y = \dfrac{-bx}{a} = \dfrac{-2x}{3}$ becomes:

$$\frac{\sqrt{2}}{2}(y - x) = \frac{-\sqrt{2}}{3}(y + x) \rightarrow y = \frac{x}{5}$$

Chapter 10

10.1:

$$\cos\theta = \frac{ADJ}{HYP}$$

$$\cos 50° = \frac{2}{r}$$

$$r = \frac{2}{\cos 50°} = 3.1"$$

10.2:

$$\sin\theta = \frac{OPP}{HYP}$$

$$\sin 60° = \frac{4}{r}$$

$$r = \frac{4}{\sin 60°} = 4.6"$$

10.3:

$$\cos\theta = \frac{ADJ}{HYP}$$

$$\cos \pi/3 = \cos 60° = \frac{y}{2}$$

$$y = 2\cos 60° = 1"$$

10.4:

$$\cos\theta = \frac{ADJ}{HYP}$$

$$\cos 50° = \frac{x}{3}$$

$$x = 3\cos 50° = 1.9"$$

10.5:

$$\tan\theta = \frac{OPP}{ADJ}$$

$$\tan 30° = \frac{y}{3}$$

$$y = 3\tan 30° = 1.7"$$

10.6:

$$\tan\theta = \frac{OPP}{ADJ}$$

$$\tan 40° = \frac{1}{x}$$

$$x = \frac{1}{\tan 40°} = 1.2"$$

10.7:

$$\tan\theta = \frac{OPP}{ADJ}$$

$$\tan\frac{\pi}{4} = \tan 45° = \frac{y}{2}$$

$$y = 2\tan 45° = 2"$$

10.8:

$$\sin\theta = \frac{OPP}{HYP}$$

$$\sin\frac{\pi}{6} = \sin 30° = \frac{y}{5}$$

$$y = 5\sin 30° = 2.5"$$

10.9: $\csc 30°$

$$\csc 30° = \frac{1}{\sin 30°} = \frac{1}{1/2} = 2$$

10.10: $\cot\left(\frac{\pi}{4}\right)$

$$\frac{\pi}{4}\cdot\frac{180}{\pi} = 45°$$

$$\cot 45° = \frac{1}{\tan 45°} = \frac{1}{1} = 1$$

10.11: $\sec\left(\frac{\pi}{6}\right)$

$$\frac{\pi}{6}\cdot\frac{180}{\pi} = 30°$$

$$\sec 30° = \frac{1}{\cos 30°} = \frac{1}{\sqrt{3}/2} = \frac{2}{\sqrt{3}} = \frac{2\sqrt{3}}{3}$$

10.12: $\csc 45°$

$$\csc 45° = \frac{1}{\sin 45°} = \frac{1}{\sqrt{2}/2} = \frac{2}{\sqrt{2}} = \sqrt{2}$$

10.13: $\frac{\csc\theta}{\cot\theta}$

$$\frac{\csc\theta}{\cot\theta} = \frac{1}{\sin\theta}\left[\frac{1}{1/\tan\theta}\right]$$

$$= \frac{\tan\theta}{\sin\theta} = \frac{\sin\theta}{\cos\theta}\left[\frac{1}{\sin\theta}\right]$$

$$= \frac{1}{\cos\theta} = \sec\theta$$

10.14: $\tan\theta\ \csc\theta$

$$\tan\theta\csc\theta = \frac{\sin\theta}{\cos\theta}\left[\frac{1}{\sin\theta}\right]$$

$$= \frac{1}{\cos\theta} = \sec\theta$$

10.15: $\cot\theta\sin\theta$

$$\cot\theta\sin\theta = \frac{1}{\tan\theta}[\sin\theta]$$

$$= \frac{\sin\theta}{\sin\theta/\cos\theta} = \cos\theta$$

10.16: $\cos\theta\csc\theta$

$$\cos\theta\csc\theta = \cos\theta\left[\frac{1}{\sin\theta}\right]$$

$$= \frac{\cos\theta}{\sin\theta} = \frac{1}{\sin\theta/\cos\theta}$$

$$= \frac{1}{\tan\theta} = \cot\theta$$

Chapter 11

<table>
<tr>
<td>

11.1:

$a = \sin^{-1} \dfrac{2\sqrt{3}}{4}$

$= \sin^{-1} \dfrac{\sqrt{3}}{2} = 60°$

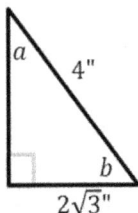

$b = \cos^{-1} \dfrac{2\sqrt{3}}{4} = \cos^{-1} \dfrac{\sqrt{3}}{2} = 30°$

Check: $90 + 60 + 30 = 180°$

</td>
<td>

11.2:

$a = \sin^{-1} \dfrac{1.5}{2}$

$= \sin^{-1} 0.75 = 48.6°$

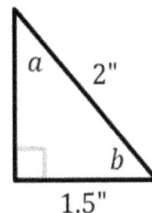

$b = \cos^{-1} \dfrac{1.5}{2} = \cos^{-1} 0.75 = 41.4°$

Check: $90 + 48.6 + 41.4 = 180°$

</td>
</tr>
<tr>
<td>

11.3:

$a = \cos^{-1} \dfrac{2}{4}$

$= \cos^{-1} \dfrac{1}{2} = 60°$

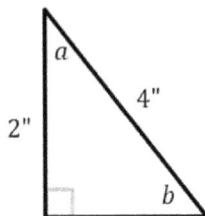

$b = \sin^{-1} \dfrac{2}{4} = \cos^{-1} \dfrac{1}{2} = 30°$

Check: $90 + 60 + 30 = 180°$

</td>
<td>

11.4:

$a = \cos^{-1} \dfrac{1}{3} = 70.5°$

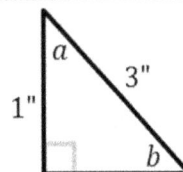

$b = \sin^{-1} \dfrac{1}{3} = 19.5°$

Check: $90 + 70.5 + 19.5 = 180°$

</td>
</tr>
<tr>
<td>

11.5:

$a = \tan^{-1} \dfrac{2}{2}$

$= \tan^{-1} 1 = 40°$

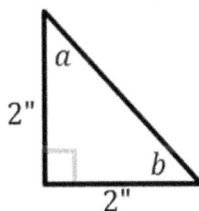

$b = \tan^{-1} \dfrac{2}{2} = 40°$

Check: $90 + 45 + 45 = 180°$

</td>
<td>

11.6:

$a = \tan^{-1} \dfrac{1}{3} = 18.4°$

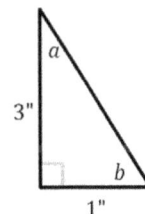

$b = \tan^{-1} \dfrac{3}{1} = 71.6°$

Check: $90 + 18.4 + 71.6 = 180°$

</td>
</tr>
</table>

11.7: $\sin^{-1}\frac{\sqrt{2}}{2}$	**11.8:** $\cos^{-1}\left(-\frac{\sqrt{3}}{2}\right)$
$\sin^{-1}\frac{\sqrt{2}}{2} = 45°$;	$\cos^{-1}\left(\frac{\sqrt{3}}{2}\right) = 30°$;
Sine values $\geq 0 \rightarrow QI, QII$	Cosine values $< 0 \rightarrow QII, QIII$
Answers are:	Answers are:
$$45° \text{ and}$$ $$180 - 45 = 135°$$	$$180 + 30 = 210° \text{ and}$$ $$180 - 30 = 150°$$
11.9: $\sin^{-1}0$	**11.10:** $\tan^{-1}(-\sqrt{3})$
$\sin^{-1}0 = 0°$:	$\tan^{-1}\sqrt{3} = 60°$:
Sine values $\geq 0 \rightarrow QI, QII$	Tangent values $< 0 \rightarrow QII, QIV$
Answers are :	Answers are:
$$0° \text{ and}$$ $$180 - 0 = 180°$$	$$-60° \text{ (or } 300°\text{) and}$$ $$180 - 60 = 120°$$
11.11: $2\sin x + 1 = 3$	**11.12:** $2\cos 2x = 1$
Let $A = \sin x$:	Let $A = \cos 2x$:
$$2A + 1 = 3$$	$$2A = 1$$
$$A = \sin x = 1$$	$$A = \cos 2x = 1/2$$
$$x = \sin^{-1}1 = 90° = \frac{\pi}{2} = 1.57$$	$$2x = \cos^{-1}\frac{1}{2} = \{\pm 60°\} = \left\{\pm\frac{\pi}{3}\right\}$$
Reviewing the plots on page 63 confirms that there is only one answer to this equation.	$$x = \left\{\pm\frac{\pi}{6}\right\} = \{\pm 0.52\}$$

11.13: $\sin^2 x = 5 - 4\sin x$

Let $A = \sin x$:

$$A^2 + 4A - 5 = 0$$

$$(A + 5)(A - 1) = 0$$

$$A = \sin x = \{-5, 1\}$$

$$\sin^{-1}(-5) \text{ does not exist}$$

$$x = \sin^{-1} 1 = 90° = \frac{\pi}{2} = 1.57$$

11.14: $-\tan^2 x = 2\tan x - 3$

Let $A = \tan x$:

$$A^2 + 2A - 3 = 0$$

$$(A + 3)(A - 1) = 0$$

$$A = \tan x = \{-3, 1\}$$

$$x = \tan^{-1}(-3) = \{108.4°, -71.6°\}$$

$$= \{0.6\pi, 0.4\pi\}$$

$$x = \tan^{-1} 1 = \{45°, 225°\}$$

$$= \left\{\frac{\pi}{4}, \frac{5\pi}{4}\right\} = \{0.79, 3.9\}$$

Note that there are four solutions to this equation.

11.15: $\cos^2 x + \frac{7}{2}\cos x = 2$

Let $A = \cos x$:

$$2A^2 + 7A - 4 = 0$$

$$(A + 4)(2A - 1) = 0$$

$$A = \cos x = \left\{-4, \frac{1}{2}\right\}$$

$$\cos^{-1}(-4) \text{ does not exist}$$

$$x = \cos^{-1}\frac{1}{2} = \{\pm 60°\} = \left\{\pm\frac{\pi}{3}\right\}$$

$$= \{\pm 1.05\}$$

11.16: $\tan^2 x = 1 + 2\tan x$

Let $A = \tan x$:

$$A^2 + 2A - 1 = 0$$

$$A = \tan x = \frac{-2 \pm \sqrt{4+4}}{2} = \{0.41, -2.41\}$$

$$\sin^{-1}(-2.41) \text{ does not exist}$$

$$x = \sin^{-1} 0.41 = \{24.5°, 155.5°\}$$

$$= \{0.14\pi, 0.86\pi\} = \{0.43, 2.71\}$$

Note: The number of decimal places you use in each step may make your answers differ slightly from mine.

Chapter 12

12.1: $\cos -45°$ $$\cos -45° = \cos 45° = \frac{\sqrt{2}}{2}$$	**12.2:** $\tan -60°$ $$\tan -60° = -\tan 60° = -\sqrt{3}$$
12.3: $\sin 135°$ $$\sin 135° = \sin(90° + 45°)$$ $$= \sin 90° \cos 45° + \sin 45° \cos 90°$$ $$= 1 \cdot \frac{\sqrt{2}}{2} + \frac{\sqrt{2}}{2} \cdot 0 = \frac{\sqrt{2}}{2}$$	**12.4:** $\cos 120°$ $$\cos 120° = \cos(2 \times 60°)$$ $$= \cos^2 60° - \sin^2 60°$$ $$= \left(\frac{1}{2}\right)^2 - \left(\frac{\sqrt{3}}{2}\right)^2 = \frac{1}{4} - \frac{3}{4} = -\frac{1}{2}$$ You get the same answer if you use $\cos 120° = \cos(90° + 30°)$.
12.5: $\cos(A - B)$ $\quad = \cos A \cos B + \sin A \sin B$ $$\cos(A + (-B))$$ $$= \cos A \cos(-B) - \sin A \sin(-B)$$ $$= \cos A \cos B - \sin A (-\sin B)$$ $$= \cos A \cos B + \sin A \sin B$$	**12.6:** $\cos 2\theta = \cos^2 \theta - \sin^2 \theta$ $$\cos 2\theta = \cos(\theta + \theta)$$ $$= \cos \theta \cos \theta - \sin \theta \sin \theta$$ $$= \cos^2 \theta - \sin^2 \theta$$
12.7: $\tan 2\theta = \frac{2 \tan \theta}{1 - \tan^2 \theta}$ $$\tan 2\theta = \tan(\theta + \theta)$$ $$= \frac{\tan \theta + \tan \theta}{1 - \tan \theta \tan \theta} = \frac{2 \tan \theta}{1 - \tan^2 \theta}$$	**12.8:** $\sin \frac{\theta}{2} = \pm\sqrt{\frac{1 - \cos \theta}{2}}$ $$\cos \theta = \cos\left(2 \cdot \frac{\theta}{2}\right) = 1 - 2 \sin^2 \frac{\theta}{2}$$ $$\sin^2 \frac{\theta}{2} = \frac{1 - \cos \theta}{2}$$ $$\sin \frac{\theta}{2} = \pm\sqrt{\frac{1 - \cos \theta}{2}}$$

12.9: $\cos A \cos B$

$\quad = \frac{1}{2}[\cos(A + B) + \cos(A - B)]$

Working from the right side:

$\quad \frac{1}{2}[\cos(A + B) + \cos(A - B)]$

$\quad = \frac{1}{2}[\cos A \cos B - \sin A \sin B$

$\quad + \cos A \cos B + \sin A \sin B\,]$

$\quad = \frac{1}{2}[2\cos A \cos B] = \cos A \cos B$

12.10: $\sin A \sin B$

$\quad = \frac{1}{2}[\cos(A - B) - \cos(A + B)]$

Working from the right side:

$\quad \frac{1}{2}[\cos(A - B) - \cos(A + B)]$

$\quad = \frac{1}{2}[\cos A \cos B + \sin A \sin B$

$\quad -(\cos A \cos B - \sin A \sin B\,]$

$\quad = \frac{1}{2}[2\sin A \sin B] = \sin A \sin B$

12.11: $\sin A - \sin B$

$\quad = 2\sin\left(\frac{A-B}{2}\right)\cos\left(\frac{A+B}{2}\right)$

Let $A' = \frac{A-B}{2}$; $\ B' = \frac{A+B}{2}$:

$\quad\quad 2\sin A' \cos B'$

$= 2\left[\frac{1}{2}[\sin(A' + B') + \sin(A' - B')]\right]$

$\quad = \sin A + \sin(-B) = \sin A - \sin B$

12.12: $\cos A + \cos B$

$\quad = 2\cos\left(\frac{A+B}{2}\right)\cos\left(\frac{A-B}{2}\right)$

Let $A' = \frac{A+B}{2}$; $\ B' = \frac{A-B}{2}$:

$\quad\quad 2\cos A' \cos B'$

$= 2\left[\frac{1}{2}[\cos(A' + B') + \cos(A' - B')]\right]$

$\quad\quad = \cos A + \cos B$

12.13: $2 \sin x \cos x = 1$

Use the identities to get the same function with the same argument:

Since $\sin(2\theta) = 2 \sin \theta \cos \theta$:

$$\sin 2x = 1$$

$$2x = \sin^{-1} 1 = \frac{\pi}{2}$$

$$x = \frac{\pi}{4}$$

12.14: $4x + 2 \sin^2 \frac{x}{2} + \cos x = 3$

Use the identities to get the same function with the same argument:

Since $\sin \frac{\theta}{2} = \pm \sqrt{\frac{1-\cos \theta}{2}}$:

$$4x + 2 \left[\frac{1-\cos x}{2} \right] + \cos x = 3$$

$$4x - \cos x + \cos x = 2$$

$$x = \frac{2}{4} = \frac{1}{2}$$

12.15: $3 \cos x = 4\sin^2 x - 3$

Use the identities to get the same function with the same argument:

Since $\sin^2 \theta + \cos^2 \theta = 1$:

$$3 \cos x = 4[1 - \cos^2 x] - 3$$

$$4 \cos^2 x + 3 \cos x - 1 = 0$$

Let $A = \cos x$:

$$4A^2 + 3A - 1 = 0$$

$$(4A - 1)(A + 1) = 0 \rightarrow A = \left\{ \frac{1}{4}, -1 \right\}$$

$$x = \cos^{-1} \frac{1}{4} = \{1.32, 4.46\}$$

And

$$x = \cos^{-1}(-1) = \pi = 3.14$$

12.16: $\sin x = \cos 2x$

Use the identities to get the same function with the same argument:

Since $\cos(2\theta) = 1 - 2 \sin^2 \theta$:

$$\sin x = 1 - 2 \sin^2 x$$

$$2 \sin^2 x + \sin x - 1 = 0$$

Let $A = \sin x$:

$$2A^2 + A - 1 = 0$$

$$(2A - 1)(A + 1) = 0 \rightarrow A = \left\{ \frac{1}{2}, -1 \right\}$$

$$x = \sin^{-1} \frac{1}{2} = \left\{ \frac{\pi}{6}, \frac{5\pi}{6} \right\}$$

And

$$x = \sin^{-1}(-1) = -\frac{\pi}{2}$$

Chapter 13

13.1:

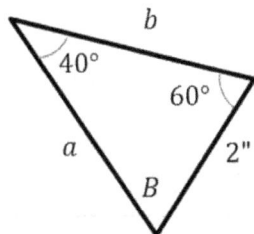

Sum of triangle angles gives:
$$B = 180 - 40 - 60 = 80°$$

AAS → Law of Sines:
$$\frac{\sin 60°}{a} = \frac{\sin 80°}{b} = \frac{\sin 40°}{2} = 0.32$$

$$a = \frac{\sin 60°}{0.32} = 2.7"; \quad b = \frac{\sin 80°}{0.32} = 3.1"$$

13.2:

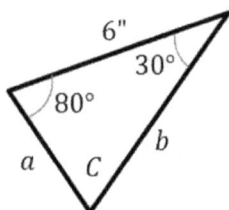

Sum of triangle angles gives:
$$C = 180 - 80 - 30 = 70°$$

ASA → Law of Sines:
$$\frac{\sin 30°}{a} = \frac{\sin 80°}{b} = \frac{\sin 70°}{6} = 0.16$$

$$a = \frac{\sin 30°}{0.16} = 3.1"; \quad b = \frac{\sin 80°}{0.16} = 6.2"$$

13.3:

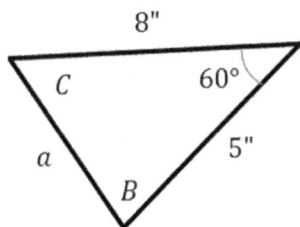

SAS → Law of Cosines:
$$a^2 = 8^2 + 5^2 - 2(40)\cos 60° \rightarrow a = 7"$$
$$8^2 = 7^2 + 5^2 - 2(5)(7)\cos B$$
$$B = \cos^{-1}(0.14) = 81.8°$$

Sum of triangle angles gives:
$$C = 180 - 60 - 81.8 = 38.2°$$

13.4:

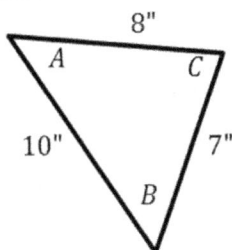

SSS → Law of Cosines:
$$10^2 = 8^2 + 7^2 - 2(56)\cos C$$
$$C = \cos^{-1}(0.12) = 83.3°$$

$$\frac{\sin A}{7} = \frac{\sin 83.3°}{10} = 0.1 \rightarrow A = \sin^{-1} 0.7 = 44.0°$$

Sum of triangle angles gives:
$$B = 180 - 44 - 83.3 = 52.7°$$

13.5:

SSS → Law of Cosines:

$$12^2 = 3^2 + 10^2 - 2(30)\cos C$$

$$C = \cos^{-1}(-0.58) = 125.7°$$

$$\frac{\sin A}{10} = \frac{\sin 125.7°}{12} = 0.07 \rightarrow A = \sin^{-1} 0.7 = 42.6°$$

Sum of triangle angles gives:

$$B = 180 - 125.7 - 42.6 = 11.7°$$

13.6:

Sum of triangle angles gives:

$$B = 180 - 80 - 30 = 70°$$

AAS → Law of Sines:

$$\frac{\sin 80°}{a} = \frac{\sin 70°}{b} = \frac{\sin 30°}{7} = 0.07$$

$$a = \frac{\sin 80°}{0.07} = 13.8"; \quad b = \frac{\sin 70°}{0.07} = 13.2"$$

13.7:

SSA → Law of Cosines:

$$10^2 = a^2 + 6^2 - 12a\cos 100° \rightarrow a^2 - 2.1a - 64 = 0$$

$$a = \frac{2.1 \pm \sqrt{4.4 + 256}}{2} = 9.1"$$

$$\frac{\sin A}{9.1} = \frac{\sin 100°}{10} = 0.1 \rightarrow A = \sin^{-1} 0.9 = 63.7°$$

Sum of triangle angles gives:

$$B = 180 - 63.7 - 100 = 16.3°$$

13.8:

SAS → Law of Cosines:

$$a^2 = 2^2 + 3^2 - 2(6)\cos 65° \rightarrow a = 2.8"$$

$$2^2 = 2.8^2 + 3^2 - 2(2.8)(3)\cos B$$

$$B = \cos^{-1}(0.76) = 40.2°$$

Sum of triangle angles gives:

$$C = 180 - 65 - 40.2 = 74.8°$$

13.9:

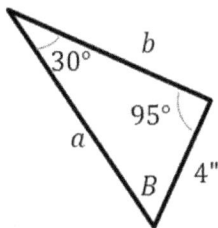

Sum of triangle angles gives:

$$B = 180 - 95 - 30 = 55°$$

AAS → Law of Sines:

$$\frac{\sin 95°}{a} = \frac{\sin 55°}{b} = \frac{\sin 30°}{4} = 0.125$$

$$a = \frac{\sin 95°}{0.125} = 8"; \quad b = \frac{\sin 55°}{0.125} = 6.6"$$

13.10:

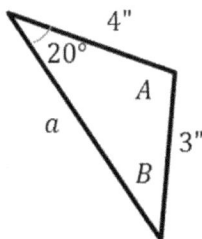

SSA → Law of Cosines:

$$3^2 = a^2 + 4^2 - 8a \cos 20° \rightarrow a^2 - 7.5a + 7 = 0$$

$$a = \frac{7.5 \pm \sqrt{56.3 - 28}}{2} = \{1.1, 6.4\} \text{ Choose 6.4" as per figure.}$$

$$\frac{\sin A}{6.4} = \frac{\sin 20°}{3} = 0.11 \rightarrow A = \sin^{-1} 0.73 = 47°$$

However, A looks obtuse, so the angle is QII or 133°

$$B = 180 - 133 - 20 = 27°$$

13.11:

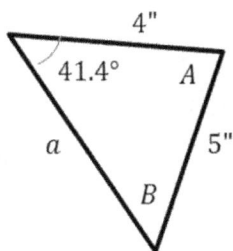

SSA → Law of Cosines:

$$5^2 = a^2 + 4^2 - 8a \cos 41.4° \rightarrow a^2 - 6a - 9 = 0$$

$$a = \frac{6 \pm \sqrt{36 + 36}}{2} = 7.2"$$

$$\frac{\sin A}{7.2} = \frac{\sin 41.4°}{5} = 0.13 \rightarrow A = \sin^{-1} 0.95 = 73.3°$$

$$B = 180 - 41.1 - 73.3 = 65.3°$$

13.12:

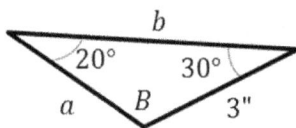

Sum of triangle angles gives:

$$B = 180 - 20 - 30 = 130°$$

AAS → Law of Sines:

$$\frac{\sin 30°}{a} = \frac{\sin 130°}{b} = \frac{\sin 20°}{3} = 0.114$$

$$a = \frac{\sin 30°}{0.114} = 4.4"; \quad b = \frac{\sin 130°}{0.114} = 6.7"$$

13.13:

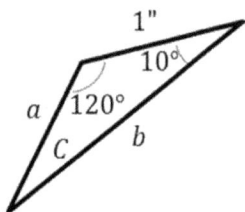

Sum of triangle angles gives:

$$C = 180 - 120 - 10 = 50°$$

ASA → Law of Sines:

$$\frac{\sin 10°}{a} = \frac{\sin 120°}{b} = \frac{\sin 50°}{1} = 0.766$$

$$a = \frac{\sin 10°}{0.766} = 0.23"; \quad b = \frac{\sin 120°}{0.766} = 1.13"$$

13.14:

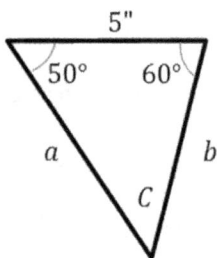

Sum of triangle angles gives:

$$C = 180 - 50 - 60 = 70°$$

ASA → Law of Sines:

$$\frac{\sin 60°}{a} = \frac{\sin 50°}{b} = \frac{\sin 70°}{5} = 0.188$$

$$a = \frac{\sin 60°}{0.188} = 4.6"; \quad b = \frac{\sin 50°}{0.188} = 4.1"$$

13.15:

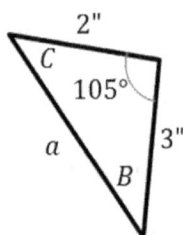

SAS → Law of Cosines:

$$a^2 = 2^2 + 3^2 - 2(6)\cos 105° \ \to a = 4"$$

$$2^2 = 4^2 + 3^2 - 2(4)(3)\cos B$$

$$B = \cos^{-1}(0.88) = 29°$$

Sum of triangle angles gives:

$$C = 180 - 105 - 29 = 46°$$

13.16:

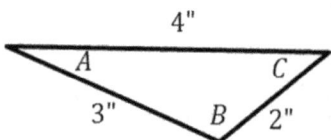

SSS → Law of Cosines:

$$3^2 = 2^2 + 4^2 - 2(8)\cos C$$

$$C = \cos^{-1}(0.69) = 46.6°$$

$$\frac{\sin A}{2} = \frac{\sin 46.6°}{3} = 0.24 \to A = \sin^{-1} 0.48 = 29°$$

Sum of triangle angles gives:

$$B = 180 - 29 - 46.6 = 104.4°$$

Chapter 14

14.1: $\sinh(-\theta) = -\sinh\theta$	**14.2:** $\cosh(-\theta) = \cosh\theta$

14.1: $\sinh(-\theta) = -\sinh\theta$

$$\sinh(-\theta) = \frac{1}{2}\left[e^{-\theta} - e^{+\theta}\right]$$

$$= -\frac{1}{2}\left[e^{\theta} - e^{-\theta}\right] = -\sinh\theta$$

14.2: $\cosh(-\theta) = \cosh\theta$

$$\cosh(-\theta) = \frac{1}{2}\left[e^{-\theta} + e^{+\theta}\right]$$

$$= \frac{1}{2}\left[e^{\theta} + e^{-\theta}\right] = \cosh\theta$$

14.3: $\cosh\theta + \sinh\theta = e^{\theta}$

$$\frac{1}{2}\left[e^{\theta} + e^{-\theta} + e^{\theta} - e^{-\theta}\right]$$

$$= \frac{2}{2}e^{\theta} = e^{\theta}$$

14.4: $\cosh\theta - \sinh\theta = e^{-\theta}$

$$\frac{1}{2}\left[e^{\theta} + e^{-\theta} - (e^{\theta} - e^{-\theta})\right]$$

$$= \frac{2}{2}e^{-\theta} = e^{-\theta}$$

14.5: $\sinh(A + B) = \sinh A \cosh B + \sinh B \cosh A$

$$\sinh(A + B) = \frac{1}{2}\left[e^{A+B} - e^{-(A+B)}\right] = \frac{1}{2}\left[e^A e^B - e^{-A}e^{-B}\right]$$

$$= \frac{1}{2}\left[(\cosh A + \sinh A)(\cosh B + \sinh B) - (\cosh A - \sinh A)(\cosh B - \sinh B)\right]$$

$$= \frac{1}{2}\left[(\cosh A \cosh B + \cosh A \sinh B + \sinh A \cosh B + \sinh A \sinh B)\right.$$
$$\left. -(\cosh A \cosh B - \cosh A \sinh B - \sinh A \cosh B + \sinh A \sinh B)\right]$$

$$= \frac{1}{2}\left[2\cosh A \sinh B + \sinh A \cosh B\right] = \sinh A \cosh B + \sinh B \cosh A$$

14.6: $\cosh(A + B) = \cosh A \cosh B + \sinh A \sinh B$

$$\cosh(A + B) = \frac{1}{2}\left[e^{A+B} + e^{-(A+B)}\right] = \frac{1}{2}\left[e^A e^B + e^{-A}e^{-B}\right]$$

$$= \frac{1}{2}\left[(\cosh A + \sinh A)(\cosh B + \sinh B) + (\cosh A - \sinh A)(\cosh B - \sinh B)\right]$$

$$= \frac{1}{2}\left[(\cosh A \cosh B + \cosh A \sinh B + \sinh A \cosh B + \sinh A \sinh B)\right.$$
$$\left. +(\cosh A \cosh B - \cosh A \sinh B - \sinh A \cosh B + \sinh A \sinh B)\right]$$

$$= \frac{1}{2}\left[2\cosh A \cosh B + \sinh A \sinh B\right] = \cosh\ A \cosh B + \sinh A \sinh B$$

14.7: $\sinh(A - B)$ $= \sinh A \cosh B - \sinh B \cosh A$ $\sinh(A - B)$ $= \sinh A \cosh(-B) + \sinh(-B) \cosh A$ $= \sinh A \cosh B - \sinh B \cosh A$	**14.8:** $\cosh(A - B)$ $= \cosh A \cosh B - \sinh A \sinh B$ $\cosh(A - B)$ $= \cosh A \cosh(-B) + \sinh A \sinh(-B)$ $= \cosh A \cosh B - \sinh A \sinh B$
14.9: $\sinh 2\theta = 2 \sinh \theta \cosh \theta$ $\sinh 2\theta = \sinh(\theta + \theta)$ $= \sinh \theta \cosh \theta + \sinh \theta \cosh \theta$ $= 2 \sinh \theta \cosh \theta$	**14.10:** $\cosh 2\theta$ $= \cosh^2 \theta + \sinh^2 \theta$ $= 1 + 2 \sinh^2 \theta = 2 \cosh^2 \theta - 1$ $\cosh 2\theta = \cosh(\theta + \theta)$ $= \cosh \theta \cosh \theta + \sinh \theta \sinh \theta$ $= \cosh^2 \theta + \sinh^2 \theta$ Using $\cosh^2 \theta = 1 + \sinh^2 \theta$ gives us the other relations.
14.11: $\cosh \frac{\theta}{2} = \pm \sqrt{\frac{\cosh \theta + 1}{2}}$ $\cosh \theta = 2 \cosh^2 \frac{\theta}{2} - 1$ $\cosh \frac{\theta}{2} = \pm \sqrt{\frac{\cosh \theta + 1}{2}}$	**14.12:** $\sinh \frac{\theta}{2} = \pm \sqrt{\frac{\cosh \theta - 1}{2}}$ $\cosh \theta = 1 + 2 \sinh^2 \frac{\theta}{2}$ $\sinh \frac{\theta}{2} = \pm \sqrt{\frac{\cosh \theta - 1}{2}}$

14.13: $\sinh A \cosh B$
$$= \frac{1}{2}[\sinh(A+B) + \sinh(A-B)]$$

$$\sinh A \cosh B = \frac{1}{4}[e^A - e^{-A}][e^B + e^{-B}]$$

$$= \frac{1}{4}[e^{A+B} + e^{A-B} - e^{B-A} - e^{-(A+B)}]$$

$$= \frac{1}{4}[e^{A+B} - e^{-(A+B)} + e^{A-B} - e^{-(A-B)}]$$

$$= \frac{1}{4}[2\sinh(A+B) + 2\sinh(A-B)]$$

$$= \frac{1}{2}[\sinh(A+B) + \sinh(A-B)]$$

14.14: $\cosh A \cosh B$
$$= \frac{1}{2}[\cosh(A+B) + \cosh(A-B)]$$

$$\cosh A \cosh B = \frac{1}{4}[e^A + e^{-A}][e^B + e^{-B}]$$

$$= \frac{1}{4}[e^{A+B} + e^{A-B} + e^{B-A} + e^{-(A+B)}]$$

$$= \frac{1}{4}[e^{A+B} + e^{-(A+B)} + e^{A-B} + e^{-(A-B)}]$$

$$= \frac{1}{4}[2\cosh(A+B) + 2\cosh(A-B)]$$

$$= \frac{1}{2}[\cosh(A+B) + \cosh(A-B)]$$

14.15: $2\sinh\frac{A+B}{2}\cosh\frac{A-B}{2} = \sinh A + \sinh B$

$$2\sinh\frac{A+B}{2}\cosh\frac{A-B}{2} = \frac{2}{2}\left[\sinh\left[\frac{(A+B)}{2} + \frac{A-B}{2}\right] + \sinh\left[\frac{(A+B)}{2} - \frac{A-B}{2}\right]\right]$$

$$= \sinh A + \sinh B$$

14.16: $2\cosh\frac{A+B}{2}\cosh\frac{A-B}{2} = \cosh A + \cosh B$

$$2\cosh\frac{A+B}{2}\cosh\frac{A-B}{2} = \frac{2}{2}\left[\cosh\left[\frac{(A+B)}{2} + \frac{A-B}{2}\right] + \cosh\left[\frac{(A+B)}{2} - \frac{A-B}{2}\right]\right]$$

$$= \cosh A + \cosh B$$

Appendix D: Trigonometric Identities

Inverse:

$$\csc\theta = \frac{1}{\sin\theta} \qquad \sec\theta = \frac{1}{\cos\theta} \qquad \cot\theta = \frac{1}{\tan\theta}$$

$$\sin\theta = \frac{1}{\csc\theta} \qquad \cos\theta = \frac{1}{\sec\theta} \qquad \tan\theta = \frac{1}{\cot\theta}$$

Negative Angle:

$$\sin-\theta = -\sin\theta \qquad \cos-\theta = \cos\theta \qquad \tan-\theta = -\tan\theta$$
$$\csc-\theta = -\csc\theta \qquad \sec-\theta = \sec\theta \qquad \cot-\theta = -\cot\theta$$

Pythagorean:

$$\sin^2\theta + \cos^2\theta = 1 \qquad 1 + \tan^2\theta = \sec^2\theta \qquad 1 + \cot^2\theta = \csc^2\theta$$

Angle Sum & Difference:

$$\sin(A \pm B) = \sin A \cos B \pm \sin B \cos A$$

$$\cos(A \pm B) = \cos A \cos B \mp \sin A \sin B$$

$$\tan(A \pm B) = \frac{\tan A \pm \tan B}{1 \mp \tan A \tan B}$$

Double Angle:

$$\sin(2\theta) = 2\sin\theta\cos\theta$$

$$\cos(2\theta) = \cos^2\theta - \sin^2\theta =$$

$$2\cos^2\theta - 1 = 1 - 2\sin^2\theta$$

$$\tan(2\theta) = \frac{2\tan\theta}{1-\tan^2\theta}$$

Half Angle:

$$\sin\frac{\theta}{2} = \pm\sqrt{\frac{1-\cos\theta}{2}} \qquad \cos\frac{\theta}{2} = \pm\sqrt{\frac{1+\cos\theta}{2}} \qquad \tan\frac{\theta}{2} = \pm\sqrt{\frac{1-\cos\theta}{1+\cos\theta}}$$

Product:

$$\sin A \cos B =$$
$$\frac{1}{2}[\sin(A + B) + \sin(A - B)]$$

$$\cos A \cos B =$$
$$\frac{1}{2}[\cos(A + B) + \cos(A - B)]$$

$$\sin A \sin B =$$
$$\frac{1}{2}[\cos(A - B) - \cos(A + B)]$$

Factoring:

$$\sin A \pm \sin B =$$
$$2\sin\left(\frac{A\pm B}{2}\right)\cos\left(\frac{A\mp B}{2}\right)$$

$$\cos A + \cos B =$$
$$2\cos\left(\frac{A+B}{2}\right)\cos\left(\frac{A-B}{2}\right)$$

$$\cos A - \cos B =$$
$$-2\sin\left(\frac{A+B}{2}\right)\sin\left(\frac{A-B}{2}\right)$$

Appendix E: Hyperbolic Trig Identities

Inverse:

$$\operatorname{csch}\theta = \frac{1}{\sinh\theta}$$

$$\operatorname{sehc}\theta = \frac{1}{\cosh\theta}$$

$$\coth\theta = \frac{1}{\tanh\theta}$$

$$\sinh\theta = \frac{1}{\operatorname{csch}\theta}$$

$$\cosh\theta = \frac{1}{\operatorname{sech}\theta}$$

$$\tanh\theta = \frac{1}{\coth\theta}$$

Negative Angle:

$$\sinh-\theta = -\sinh\theta$$
$$\operatorname{csch}-\theta = -\operatorname{csch}\theta$$

$$\cosh-\theta = \cosh\theta$$
$$\operatorname{sech}-\theta = \operatorname{sech}\theta$$

$$\tanh-\theta = -\tanh\theta$$
$$\coth-\theta = -\coth\theta$$

Pythagorean:

$$\cosh^2\theta - \sinh^2\theta = 1 \qquad \tanh^2\theta + \operatorname{sech}^2\theta = 1 \qquad \coth^2\theta - \operatorname{csch}^2\theta = 1$$

Angle Sum & Difference:

$$\sinh(A \pm B) = \sinh A \cosh B \pm \sinh B \cosh A$$

$$\cosh(A \pm B) = \cosh A \cosh B \pm \sinh A \sinh B$$

$$\tanh(A \pm B) = \frac{\tanh A \pm \tanh B}{1 \pm \tanh A \tanh B}$$

Double Angle:

$$\sinh(2\theta) = 2\sinh\theta\cosh\theta$$

$$\cosh(2\theta) = \cosh^2\theta + \sinh^2\theta =$$

$$2\cosh^2\theta - 1 = 1 + 2\sinh^2\theta$$

$$\tanh(2\theta) = \frac{2\tanh\theta}{1 + \tanh^2\theta}$$

Half Angle:

$$\sinh\frac{\theta}{2} = \pm\sqrt{\frac{\cosh\theta - 1}{2}} \qquad \cosh\frac{\theta}{2} = \pm\sqrt{\frac{\cosh\theta + 1}{2}} \qquad \tanh\frac{\theta}{2} = \pm\sqrt{\frac{\cosh\theta - 1}{\cosh\theta + 1}}$$

Product:

$$\sinh A \cosh B =$$
$$\frac{1}{2}[\sinh(A+B) + \sinh(A-B)]$$

$$\cosh A \cosh B =$$
$$\frac{1}{2}[\cosh(A+B) + \cosh(A-B)]$$

$$\sinh A \sinh B =$$
$$\frac{1}{2}[\cosh(A+B) - \cosh(A-B)]$$

Factoring:

$$\sinh A \pm \sinh B =$$
$$2\sinh\left(\frac{A\pm B}{2}\right)\cosh\left(\frac{A\mp B}{2}\right)$$

$$\cosh A + \cosh B =$$
$$2\cosh\left(\frac{A+B}{2}\right)\cosh\left(\frac{A-B}{2}\right)$$

$$\cosh A - \cosh B =$$
$$2\sinh\left(\frac{A+B}{2}\right)\sinh\left(\frac{A-B}{2}\right)$$

Index

"0" exponent	9	binomials	11
		branch	44

A-B

abscissa	29	**C**	
absolute value with exponents	9	center	39
		circle	39
addition		ellipse	41
of complex numbers	14	hyperbola	43
of exponents	9		
of fractions	8	circle	
		definition	39
additive identity	10	unit	71
adjacent side	61	circle & conic summary	45
angle difference identity	73	circular trigonometric function	82
angle sum identity	73	combination	48
arc-functions	67	commutative property	10
associative property	10	complex conjugate	15
asymptote	44	complex number	14
attenuation constant	31	complimentary angles	61
base		composite function	
exponential	9, 21	function	18
logarithm	21	numbers	7

congruent transformations	51
conic sections, conics	44
conjugate axis	43
cos, cosine	63
cosh, hyperbolic cosine	82
cot, cotangent	65
coth, hyperbolic cotangent	83
csc, cosecant	65
csch, hyperbolic cosecant	83
cubic expression	30

D

decay	31
denominator	8
dependent equations	47
equations	47
variable	29
diameter	39

directrix	40
distance	33
distributive property	10
division	
of complex numbers	15
of exponents	9
of fractions	8
domain	29
double angle identity	74

E-F

e	22
eccentricity	42
ellipse	41
Euler's identity	81
exponent	9
exponential	21
exponential graph	31
factorial	25

factoring identity — 75

focus
  ellipse — 41
  hyperbola — 43
  parabola — 40

FOIL — 11

fractional exponents — 9

fractions — 8

function — 17

G-H

graph — 29

graphing solutions — 50

half angle identity — 75

horizontal line — 36

hyperbola — 43

hyperbolic trig function — 82

hyperbolic trig identities (abbr.) — 82, 83

hypotenuse — 61

I-K

i — 13

identity
  additive — 10
  angle difference — 73
  angle sum — 73
  double angle — 74
  factoring — 75
  half angle — 75
  hyperbolic trig (abbr.) — 82, 83
  inverse — 71
  multiplicative — 10
  negative angle — 71
  product — 75
  property — 10
  Pythagorean — 72
  table of hyperbolic trig — 147
  table of trig — 146

imaginary operator
  operator — 14
  unit or number, i — 13

independent equations
  equations — 47
  variable — 29

integers — 7

inverse
    function 19
    hyperbolic trig functions 84
    identity 71
    log or exponential 21
    property 10
    trigonometric functions 67

irrational numbers 7

L-M

law of
    cosines 79
    sines 77

least common denominator (LCD) 8

linear expression 30

ln 22

logarithm, log
    base conversion 23
    graph 31
    log 21

major axis, ellipse 41

midpoint 33

minor axis 41

multiplication
    of complex numbers 14
    of exponents 9
    of fractions 8

multiplicative identity 10

N-O

natural exponential, e 22

natural logarithm, ln 22

negative angle identity 71

negative
    integer exponents 9
    numbers 7

numerator 8

opposite side 61

order
    of operations 10
    of polynomial 15

ordinate 29

origin 29

P-Q			R-S	
parabola	40		radical	8
parallel lines	37		radius	39
perpendicular lines	37		raising to a power	9
point-slope form	35		range	29
positive integer exponents	9		rational numbers	7
positive numbers	7		rationalization of fractions	8
power	9		real numbers	
			numbers	7
prime numbers	7		operator	14
product identity	75		reduction of fractions	8
properties			reflection about the axes	51, 54
of addition & multiplication	10			
of logarithms, exponentials	23		right triangle	61
Pythagorean identities	72		roots	15
quadrant	62		rotation	51, 54
quadratic			sec, secant	65
equation	11, 15			
expression, graphing	30		sech, hyperbolic secant	83
polynomial	15			

set notations	14	trigonometric functions		63
simultaneously solving equations	47	unit circle		71
sin, sine	61, 63	**V-W**		
sinh, hyperbolic sine	82	vertex		
		ellipse		41
slope	34	hyperbola		43
		parabola		40
slope intercept form	35			
		vertical line		36
substitution	49			
		vertical line test		17
subtraction				
of complex numbers	14	whole numbers		7
of exponents	9			
of fractions	8	**X-Z**		
systems of equations	47	x-axis		29
T-U		x-y plane		29
tan, tangent	63	y-axis		29
tanh, hyperbolic tangent	83	y-intercept		35
transformations	51			
translation	51			
transverse axis	43			

Other Books In This Series

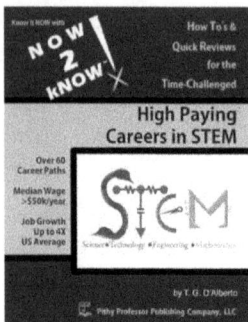

NOW 2 kNOW™ High Paying Careers in STEM
Plan for a career where your education dollars and hard work really pay off. STEM careers (science, technology, engineering, and math) are high paying, and companies are scrambling to find qualified candidates in the U.S. Better yet, the work is rewarding with opportunities to really impact the world!

NOW 2 kNOW™ Calculus I
Calculus is the gateway to many financially stable and successful careers. You might think this would make it hard and inaccessible, but that simply isn't true. Calculus is actually very easy! Once you see the concepts outlined succinctly, you'll see how little it takes to become a master.

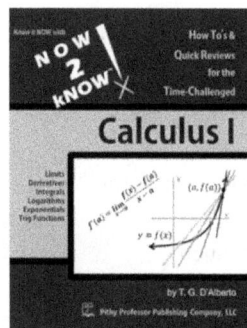

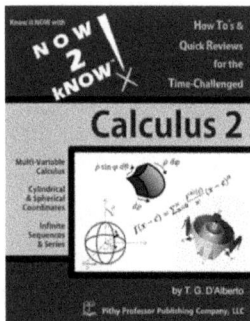

NOW 2 kNOW™ Calculus 2
Calculus 2 builds on the concepts of Calculus 1 with multi-variable functions and adds new concepts with infinite sequences and series. With thorough yet concise explanations and over 200 problems and worked out solutions, the NOW 2 kNOW™ Calculus 2 text makes learning math much easier!

NOW 2 kNOW™ Algebra I

NOW 2 kNOW™ Geometry

Need a quick review of Algebra I or Geometry to give you a stronger foundation? Thorough and concise instruction with over 200 problems and worked out solutions are in each text!

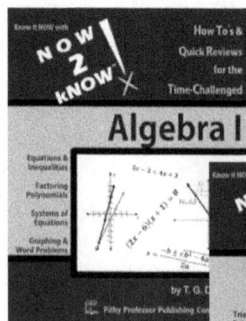

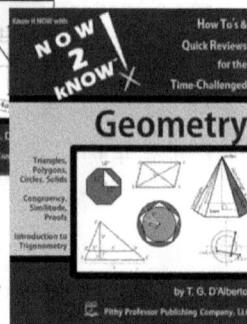

Go to Amazon.com and Search "NOW 2 kNOW"
or visit www.NOW2kNOW.com for updates.

www.ingramcontent.com/pod-product-compliance
Lightning Source LLC
Chambersburg PA
CBHW062026210326
41519CB00060B/7145